RAND NATIONAL DEFENSE RESEARCH INSTITUTE

An Approach to Life-Cycle Management of Shipboard Equipment

Bradley Martin, Roland J. Yardley, Phillip Pardue, Brynn Tannehill, Emma Westerman, Jessica Duke

Prepared for the United States Navy

For more information on this publication, visit www.rand.org/t/RR2510

Library of Congress Cataloging-in-Publication Data is available for this publication.
ISBN: 978-1-9774-0144-1

Published by the RAND Corporation, Santa Monica, Calif.
© Copyright 2018 RAND Corporation
RAND® is a registered trademark.

Cover: Hull Maintenance Technician Fireman Cherry Brown grinds metal aboard the USS Arleigh Burke
(U.S. Navy photo by Mass Communication Specialist Seaman Raymond Maddocks).

Support RAND
Make a tax-deductible charitable contribution at
www.rand.org/giving/contribute

www.rand.org

Preface

Navy ships are a complex combination of different systems and pieces of equipment, ranging from hull structure to tankage to individual components of combat systems. All these systems have maintenance, training, and supply requirements, some of which are known and some of which arise unexpectedly. Navy processes for managing these system life cycles are not efficiently organized and result in seams between ships in new construction and those already in service. This report reviews and assesses the life-cycle management of equipment installed on U.S. Navy ships, and recommends management procedures for improving the readiness of this equipment, increasing its efficiency, and reducing sustainment costs.

This research was sponsored by Program Executive Office (PEO) Ships and Naval Sea Systems Command (NAVSEA) and conducted within the Acquisition and Technology Policy Center of the RAND National Defense Research Institute, a federally funded research and development center sponsored by the Office of the Secretary of Defense, the Joint Staff, the Unified Combatant Commands, the Navy, the Marine Corps, the defense agencies, and the defense Intelligence Community.

For more information on the RAND Acquisition and Technology Policy Center, see www.rand.org/nsrd/ndri/centers/atp or contact the director (contact information is provided on the webpage).

Contents

Figures and Tables

Tables

Summary

Background and Purpose

Navy ships combine a complex array of equipment and systems, ranging from propulsion to combat systems to electronics to food preparation. Some systems are installed when the ship is under construction. Others are installed or upgraded later. Certain items are common shipboard equipment (CSE), meaning that they are used on several types of ships. These range from hulls to components of combat systems. All have different training, maintenance, and supply requirements. Navy ships also undergo a wide range of planned and unplanned events during their life cycles. Planned events extend from new construction to sustainment and training, while unplanned events include equipment failures and obsolescence.

Many organizations deal with CSE. Some handle only new construction, some conduct maintenance and sustainment of in-service ships, and some work on specific ship systems. Ideally, common processes and databases would apply to both planned and unplanned events. However, this does not appear to be the case. These organizational misalignments may be complicating the life-cycle maintenance of ships.

Such misalignments can have a number of consequences. Ships of a given class may and do have different equipment. New ships arrive with upgraded equipment, while installation of that equipment for in-service ships lags behind. Ships of different ages in the same class have different equipment, making it difficult to manage documentation, parts, and maintenance even within a single class of ships, let alone across the fleet.

This report explores the life-cycle management (LCM) of shipboard equipment It looks at empirical trends in aspects of LCM and then assesses the organizational relationships supporting the installation, use, repair, and replacement of CSE.

Approach

Our approach to the analysis had three components. First, we analyzed the trends in surface-ship LCM using the Guided Missile Destroyer (DDG)–51 class as the exemplar case. The DDG-51 was selected due to it being the most numerous surface-ship class with ships in all phases of LCM, from construction to sustainment. We next examined new construction processes to identify any challenges that might be complicating LCM. Finally, we assessed organizational issues that might be contributing to the problems we identified, focusing especially on data stovepipes.

Findings

Life-Cycle Management Trends

Based on analysis of the DDG-51 class, the Navy's most numerous surface-ship class and the one with the widest variation in ship age, we identified several LCM trends. First, ship depot maintenance availabilities grew in 2013–2016, nearly doubling in cost from 2012. Moreover, as a proportion of the total operations and sustainment (O&S) cost, ship depot maintenance grew from being a small portion of the total in 2001 to nearly a third in 2014. A significant proportion of work being done on ships in a depot maintenance period is unplanned. The classes of equipment most likely to grow in cost and scope are associated with hull structure and ballast tanks. Deferral of maintenance tends to increase the scope of the work and, inevitably, the expense, and the Navy has made deferral common.

Similarly, the systemic attention to equipment that costs a lot but might not break very frequently might detract from examination of common parts that would in fact more effectively reduce long term costs.

Second, the Navy generally does not provide a systems-level view of troubled equipment. The Naval Surface Warfare Center-Corona (NWSC-C), publishes a "troubled-systems list" including the systems that might be causing a problem, but it is a snapshot, not a long-range planning tool. Thus, it is not possible to tell whether a system is troubled because of something unusual or because some of its components have met or exceeded their design life. The troubled-system approach also tends to highlight the failures of the expensive systems. Less expensive failures receive less emphasis even if the overall cost of many low-cost failures exceed those of the high-cost systems. The casualty report (CASREP) system calls attention to equipment that limits a ship's functioning, but the reports do not reflect every equipment failure.

Third, it is not possible to develop a complete picture of the systems driving maintenance costs without consulting multiple data sources. The data systems are not connected, and the result is an ad hoc series of processes to deal with different parts of the problem, and one that does not produce a complete and accurate picture of the material state of ships in service. Our main conclusion from a review of the evidence on sustainment is that the Navy clearly makes an effort to deal with material issues as they arise. However, there is an apparent lack of focus on longer-term issues and no integrated database in which to manage them.

New Construction

The Navy has an elaborate process for building ships and delivering them to the fleet. Decisions made during that process can have downstream effects on the sustainment of CSE, and there are indeed multiple examples. Issues associated with these decisions fall into the following four categories:

- Equipment is defective because of shipyard or contractor construction issues.
- Ships are delivered with equipment that is already over budget.
- Ships are delivered with equipment that does not work because technological assumptions were overly optimistic.
- Ships are delivered with equipment that does not work because of planning or design errors.

Improving the Common Shipboard Equipment Life-Cycle Process

A common theme running through our analysis is that the problems with CSE are not caused by technical issues that could not have been anticipated or were beyond the Navy's technical capability to address. Rather, it appears that certain aspects of organization and process hamper the ability to carry out a cohesive response. We identify three areas in particular: data reporting and compatibility; funding and incentive structures; lack of a common command perspective.

With respect to data, we find it striking that so little of it is compatible. Much data is collected, but it is not easily shared. Our analysis suggests that a fundamental problem with the current approach to data management is that the existence of numerous disparate data sources makes it a challenge for stakeholders to develop a complete picture of material and other issues driving life-cycle maintenance costs.

With respect to funding, whereas construction funding extends for seven years, sustainment is resourced from annually appropriated operations and maintenance accounts. This arrangement is paradoxical in that it creates a long-term funding structure for the organization that is most focused on the short-term requirement of delivering a ship, but imposes a short-term structure on the organizations responsible for sustaining the ships through the longest portion of the life cycle. The resource and budgeting cycles create disincentives for a long-term management outlook.

Organizationally, the shipbuilding, modernization, and maintenance enterprise within Naval Sea Systems Command (NAVSEA) consists of the Program Executive Office (PEO) Ships, SEA21, and PEO Integrated Warfare Systems (IWS). The requirements of the three organizations compete for resources throughout the life-cycle of the ship. At the same time, the Type Commander of the Naval Surface Force (CNSF) is responsible for articulating the requirements while the Office of the Chief of Naval Operations (OPNAV) staff serves as the resource sponsor. One upshot of this organization is the creation of budget stovepipes in shipbuilding, modernization, and fleet maintenance.

Conclusions

Our analysis leads us to the following six conclusions:

- The Navy's "troubled-systems" approach toward immediate rather than long-term issues identifies systems after they fail in service, and tends to focus on the individual equipment/systems rather than those that are collectively more expensive.
- Similarly, the systemic attention to equipment that costs a lot but might not break very frequently might detract from examination of common parts that, if repaired or replaced, would in fact more effectively reduce long-term costs.
- The many databases and systems in use focus on specific parts of the life-cycle process and lack coherence. Taken together, these data sources could not produce a complete and accurate picture of the state of ships in service.
- Differences in command focus, amplified by differences in funding mechanisms for resourcing different aspects of the ship's life cycle, create different incentive structures.
- Reliance on the Planning, Programming, Budgeting, and Execution (PPBE) process as a means to allocate resources drives a focus on influencing the Program Objective Memorandum (POM) as opposed to defining requirements and reviewing existing data. Improvement proceeds at no faster than the pace of the PPBE process.
- No one is empowered to adjudicate among different organizational priorities.

Recommendations

We offer recommendations in four areas.

Data Sets and Systems

Data Standards

The Navy needs to generate and enforce common data standards across the whole of the enterprise by generating a data management system with the following characteristics:

- accessibility across the enterprise
- agreement that the data derived from an agreed family of systems forms the basis for decision in every aspect of ship LCM
- either a common temporal horizon or at least the ability to depict decisions across multiple such horizons
- ability to relate different aspects of the readiness framework into a common framework, specifically in order to provide understanding of how decisions made relating to one aspect of the life cycle affects some other aspect.

Essential Common Data

The following list suggests the kinds of data that need to be collected about systems and processes that span ship service life and apply to all the major elements of material readiness:

- the failure rate, by component, of installed equipment across the fleet, whether this failure is corrected by repair or replacement of the component or by replacement of the overall system in which it functions
- the specific long-range maintenance schedules for ships in service, including the requirements identified in particular availabilities as necessary by NAVSEA Technical Warrant Holders in Technical Foundations Papers or like documents
- the specific cost of deferred maintenance, possibly by historical comparison of jobs with like ship's work line-item numbers (SWLINs) modernization plans and schedules as generated by PEO Ships and PEO IWS to include the maintenance availabilities in which the modifications are planned.

The Maintenance Figure of Merit System

Maintenance Figure of Merit (MFOM) is a family of computer systems that takes inputs of maintenance data from numerous existing sources and uses models developed by subject matter experts (SMEs) to predict fleet readiness. It has the potential to enable the broader view we are recommending here, and the Navy should consider modifying it and expanding its use across the organization.

Incentive Structures

Many of the issues discussed here stem from an annual appropriations process for operations and maintenance; thus, the system would benefit from a longer-term appropriation process. Because Congress, which controls this process, has shown itself reluctant to release control over the individual accounts, the Navy cannot correct the problems caused by annual appropriations on its own. Thus, Navy leadership should make this case to Congress and within the Department of Defense. In the interim, the Navy should focus its efforts on mitigating the effects of the system, while understanding that a formal change is unlikely in the immediate future.

The Navy tends to measure success and failure in achieving staff objectives by the degree to which staff actions informed and protected a POM. Removing this as the centerpiece of process and decisionmaking may be the most concrete thing Navy leadership can do to counteract the incentives the system will generate if unchallenged.

Command and Organizational Structures

The current organizational structure for building and maintaining the surface fleet is divided primarily among four different flag officers (PEO Ships, PEO IWS, SEA21, and OPNAV N96) who report through three different reporting seniors (Assistant Secretary of the Navy, Research, Development, and Acquisition [ASN RD&A], NAVSEA, and OPNAV N9). No common superior is charged with adjudicating among the programs and proposals of these organizations. Several options exist for such a common superior. For example, it could be the Commander of NAVSEA or the Commander of Naval Surface Force (CNSF). We do not specifically endorse any particular common superior designation. We do, however, strongly recommend that one be designated, with reporting lines that make clear the need to consider the short- and long-term consequences of LCM decisions.

Response to Recent Incidents and Reviews of Surface Ship Readiness

The need for more coherent command relationships and management was identified as a surface force readiness issue in two high-level reviews conducted in response to a series of accidents involving surface ships, two of which resulted in multiple fatalities among crew members. We review in an appendix how our findings and recommendations comport with and bear on these reviews.

Acknowledgments

The authors gratefully acknowledge the sponsorship and guidance provided by RADM (ret) David Gale, RADM William Galinis, Captain Tim Crone, Mr Colin Weed, and the many dedicated members of Team Ships. We greatly appreciate the thorough and insightful reviews provided by our RAND colleagues, Tim Conley and Anthony Rosello, and our external reviewer, Doyle Hodges. The report benefited greatly from the editorial efforts of Jerry Sollinger and Barbara Bicksler. We also greatly appreciate the guidance and support of the Acquisition and Technology Policy Center leadership and support staff, including Laura Baldwin, Cynthia Cook, Chris Mouton, Leslie Thornton, and Sunny Bhatt.

Abbreviations

AIS	Automatic Identification System
Ao	operational availability
ASN RD&A	Assistant Secretary of the Navy, Research, Development, and Acquisition
ASW	auxiliary sea water
BMDRA	Ballistic Missile Defense Readiness Assessments
CASREP	casualty report
CDMD-OA	Configuration Data Managers Database—Open Architecture
CG	Guided Missile Cruiser
CNO	Chief of Naval Operations
CNSF	Commander, Naval Surface Force
CRCPT	Current Readiness Cross Pillar Team
CSE	common shipboard equipment
DDG	Guided Missile Destroyer
DMSMS	diminishing manufacturing sources and material shortage
D-SAGT	Deck Self-Assessment Groom Team
ECS	Engineering Control System
EOSS	Engineering Operational Sequencing System
ESL	expected service life
eRMS	Enterprise Remote Monitoring System
FY	fiscal year

FYDP	Future Years Defense Plan
GFE	government furnished equipment
HM&E	hull, mechanical, and electrical
INSURV	Board of Inspection and Survey
ISEA	In-Service Engineering Agent
IWS	Integrated Warfare Systems
KPP	key performance parameters
LCM	life-cycle management
LCMG	life-cycle management group
LCS	Littoral Combat Ship
LPD	landing platform dock
LRLAP	Long-Range Land Attack Projectile
LSC	large surface combatants
LSD	dock landing ship
MCS	machinery control system
MFOM	Maintenance Figure of Merit
MRDB	Material Readiness Database
NAVSEA	Naval Sea Systems Command
NAVSUP	Naval Supply Systems Command
NIIN	national item identification number
NMD	Navy Maintenance Database
NNSY	Norfolk Naval Shipyard
NSWC	Naval Surface Warfare Center
NSWC-C	Naval Surface Warfare Center-Corona
O&S	operations and sustainment
OPNAV	Office of the Chief of Naval Operations
PEO	Program Executive Office

PMS	Planned Maintenance System
POL	petroleum oil and lubricants
POM	Program Objective Memorandum
PPBE	Planning, Programming, Budgeting, and Execution
PPM	propulsion, power, machinery
RMC	regional maintenance center
S2E	sailor to engineer
SCD	Ship Change Document
SCN	Shipbuilding and Conversion, Navy
SDM	ship design manager
SEWMMC	Surface and Expeditionary Warfare Maintenance and Modernization Committee
SLE	service life extension
SME	subject matter expert
SUPSHIP	Supervisor of Shipbuilding Conversion and Repair
SURFMEPP	Surface Maintenance Engineering Planning Program
SWLIN	ship's work line-item number
TAD/TDY	temporary additional duty/temporary duty under instruction
TAVR	Technical Assistance Visit Reports
TFP	technical foundation paper
TSRA	Total Ship Readiness Assessment
TYCOM	Type Commander
USMC	United States Marine Corps
VAMOSC	Visibility and Management of Operating and Support Costs
VLS	Vertical Launching System

CHAPTER ONE

Introduction

Background and Purpose

Ships are complex systems with equipment for everything from propulsion to combat to medical care and food preparation. Some equipment is installed when the ship is first delivered; some is added later or upgraded. For the purposes of our study, we define common shipboard equipment (CSE) as equipment that is widely used and installed on Navy surface combatants and amphibious ships. Shipboard equipment refers to a wide variety of systems, ranging from hull structure and tankage to individual components of combat systems. All have maintenance, training, and supply requirements.

A ship's overall life cycle may include several planned and unplanned actions. Using the Department of Defense's (DoD's) description of system life cycle as a guide,[1] we consider planned actions in a ship's life cycle to include the following:

- new construction: the process of designing and carrying out construction on a new ship
- ship maintenance and modernization: the process of keeping equipment in working order and current with new technology
- readiness/sustainment/survey: the process of keeping the ship and equipment able to perform missions and supplied with required repair items
- availability and training: the process of operating the ship and training personnel to perform repair and diagnostics
- demilitarization/disposal: the process of concluding a ship's service life.

But not all events bearing on a ship's life cycle are planned. Unplanned events occur and require action. Some of these events include:

- obsolescence and diminishing manufacturing sources and material shortage (DMSMS)

---

[1] Defense Acquisition University, "Defense Acquisition Life Cycle Wall Chart," February 14, 2018.

- Equipment failures and safety issues (e.g., fires)
- Statutory/regulatory changes (e.g., new policies about environmental impact)
- Threats (e.g., cyber threats)

This report explores the life-cycle management (LCM) of CSE. It looks at empirical trends in aspects of LCM and then assesses the organizational relationships supporting the installation, use, repair, and replacement of common shipboard equipment.

The Organizational Challenge

Planned and unplanned events are interrelated, and ideally common processes and databases would deal with both. However, this does not appear to be the case. Multiple organizations play a role in some part of the CSE currently in the fleet. Some organizations are concerned with new construction, some with maintenance and sustainment of ships in service, some with particular ship systems. While LCM is inherently complicated, some of these organizational misalignments may be making it even more complicated.

Organizations Involved in Common Shipboard Equipment Life-Cycle Management

Two of the major players in surface-ship construction and LCM are Program Executive Office (PEO) Ships and SEA21. PEO Ships manages design and construction of new ships and ports directly to the Assistant Secretary of the Navy for Research, Development, and Acquisition (ASN RD&A). Naval Sea Systems Command (NAVSEA)–21 is a dedicated LCM organization for in-service ships and reports to the NAVSEA commander.

These organizations are in many ways closely linked, and indeed are referred to as "Team Ships." The admirals and senior executives in charge of these organizations are colocated, and it is common for the admiral serving as SEA21 to transition to PEO Ships. Unquestionably, these organizations want close cooperation and, indeed, they have collaborated with RAND for production of this study.

However, their missions differ, and these differences contribute to a disconnect between these organizations. Of greater significance, however, are the larger disconnects that may occur between these organizations and the ones that control requirements and funding in the Office of the Chief of Naval Operations (OPNAV) and operations and scheduling in the fleet staffs.

The Consequences of Disconnected Life-Cycle Management

We examine in some detail in a later chapter some of the life cycle and readiness issues the data show are present in the surface fleet. Conceptually, however, the consequences of disconnected processes are very clear. Ships of a given class may have different equipment. New ships may be delivered with upgraded equipment, while ships in service lag behind. Ships of different ages in the same class have different equipment, and it may be difficult to manage documentation, parts support, and maintenance, even within a single ship class. Additionally, lessons learned from operating ships in service might not be received by those charged with construction.

Avoidable problems recur as a result. Ideally, we posit that the cycle for learning and implementing solutions should resemble Figure 1.1. There should be a continuous feedback loop between the introduction of ships into the fleet, a stream of common data made available to all stakeholders, a decision to address identified problems, and then implementation of a class-wide solution. Problem solutions are thus proactive and systemic rather than reactive and episodic. An ideal approach would begin as a ship is introduced into the fleet (bottom left corner of Figure 1.1) and data are collected and analyzed to identify the troubled systems. These problem systems are then surfaced to the leadership, where solutions are developed, and then these solutions are integrated into the specification and engineering processes and, ultimately, introduced into ships in the fleet.

However, reality differs widely from the ideal. Many organizations engage in some parts of this cycle, such as collecting and analyzing data; other parts, such as prioritizing systems, lack a single responsible agency. The surface Navy in general meets all operational commitments, and actors in the system work hard to ensure that ships, particularly deploying ships, receive the required maintenance, modernization, and overall life-cycle support. However, there appears to be a persistent failure to antici-pate and plan. Surface-ship maintenance availabilities routinely grow in scope and

Figure 1.1
Construction-Modernization Feedback Loop

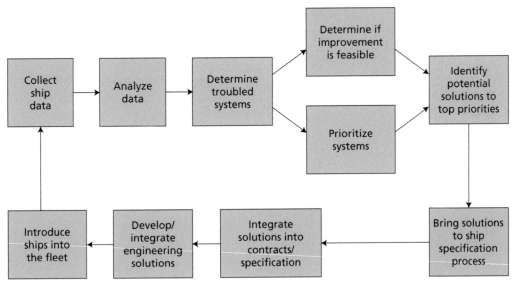

RAND RR2510NAVY-1.1

duration.[2] Modernization and maintenance efforts do not always align, although work by the Surface Maintenance Engineering Planning Program (SURFMEPP) and other agencies in the surface warfare enterprise have improved this alignment.[3] New ships are sometimes delivered with near-obsolete equipment,[4] while older ships sometimes have equipment that has been long replaced in newer ships.[5] We review these data in more detail in Chapter Two, but the need for improvement was obvious enough to PEO Ships to cause it to commission this study and share the monitoring of it with SEA21.

Report Organization

Chapter Two analyzes empirical trends in surface-ship LCM. Many of the examples displayed are from the Guided Missile Destroyer (DDG)–51 class, but the trends identified apply broadly. Chapter Three discusses some of the challenges arising in the new construction process, and how these might be complicating LCM. Chapter Four discusses in more detail the organizational issues that may be contributing to the problems identified, focusing in particular on the effect of data stovepipes. Chapter Five presents conclusions and recommendations.

---

[2]   Bradley Martin, Michael E. McMahon, James G. Kallimani, and Tim Conley, *Accounting for Growth in the Ship Depot Maintenance Account,* Santa Monica, Calif.: RAND Corporation, RR-1837-NAVY, 2017.

[3]   Robert Button, Bradley Martin, Jerry Sollinger, and Abraham Tidwell, *Assessment of Surface Ship Maintenance Requirements,* Santa Monica, Calif.: RAND Corporation, RR-1155-NAVY, 2015.

[4]   Examples include delivering new ships with Windows operating systems that were several generations out of date.

[5]   Among the examples cited for us were trash compactors that were difficult to maintain. These were replaced on new-construction ships but kept in service on older platforms at significant expense and effect on ships as they complied with environmental regulations.

CHAPTER TWO

Empirical Trends in Life-Cycle Management

We begin by examining whether there is indeed evidence of a problem in life-cycle costs of ship equipment and where these might be seen as escalating beyond what was planned. Our method for doing this is to examine different databases to determine what they can tell us about the types of systems that show cost escalation.

This task is complicated in that different databases cover different aspects of the ship's life cycle, and no consolidated database enables comparison of these different aspects. Thus, we will look at several different measurements that might provide some insight as to the various aspects of the ship life cycle.

Components of Ship Life-Cycle Cost

Figure 2.1 shows the notional components of any system's life cycle, and in particular the types of funding associated with each part:

The main focus of this report falls on the interrelationships among the different activities that are reflected in the different funding streams in the figure. A number of factors affects the cost of ship construction, ranging from technological complexity to contractor performance. These are covered in detail in other publications.[1] We are not primarily interested in these in isolation. We are instead looking at the elements of long-term sustainment that might be improved by different investments earlier in the ship's life cycle.

In terms of long-term operations and sustainment (O&S), Figure 2.2 depicts the elements of O&S costs present in Arleigh Burke DDGs, selected because they form the surface-ship class with the most ships currently in service and with a well-documented sustainment history. The purple line represents the total number of ships in the class, which currently stands at 67 ships in service. There is a predictable and proportionate increase in overall cost just based on class size. Some sustainment costs have gone up

---

[1] Mark Arena, Irv Blickstein, Obaid Younossi, and Clifford A. Grammich, *Why Has the Cost of Navy Ships Risen? A Macroscopic Examination of the Trends in U.S. Naval Ship Costs over the Past Several Decades*, Santa Monica, Calif.: RAND Corporation, MG-484-NAVY, 2006.

Figure 2.1
Components of Ship Life-Cycle Cost

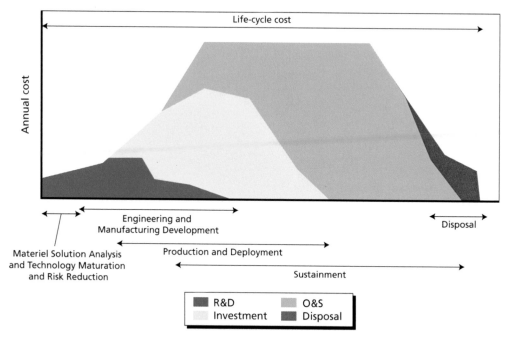

SOURCE: OSD, Cost Assessment and Program Evaluation, *O&S Cost Estimating Guide*, March 2014.
RAND *RR2510NAVY-2.1*

directly in proportion to the size of the class. Personnel costs have gone up simply as the number of ships requiring crew has increased. Fuel costs vary, but they generally stay in a narrow band. In most other components, they also vary but the increase in cost is not greater than what would be expected based on ship class size.

The two factors that appear to show some limited variation, and indeed impact on the overall size of the sustainment cost, are systems improvements (associated with modernization) and depot maintenance. The variations occur in limited numbers of years. For example, ship maintenance in 2014 nearly doubled as a proportion of overall cost from 2011, despite the addition of only two ships to the fleet. This bears further investigation. However, the existing data points do suggest that in some years ship maintenance has an outsize effect, which no other factor ever appears to exert.

Ship Depot Maintenance Costs

The data at the budget level provided in Visibility and Management of Operating and Support Costs (VAMOSC) can only suggest broad trends. These show that the overall sustainment cost of the DDG-51 class is due primarily to class size, with some suggestion that maintenance might be inducing more variation than other factors. Considerable work has recently been done on the factors that affect cost growth in ship depot

Figure 2.2
Operations and Sustainment Costs for the DDG-51 Class in Constant $FY 2018 Billions

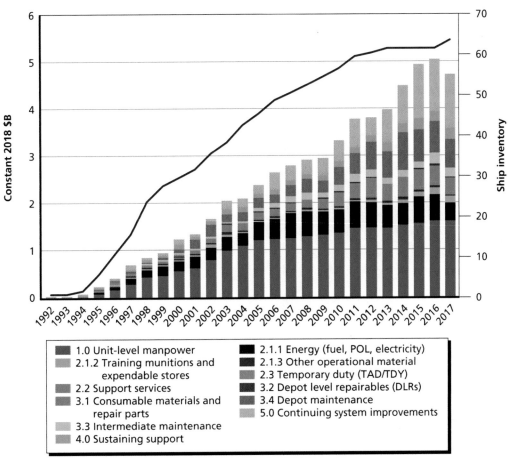

SOURCE: Navy Visibility and Management of Operating and Support Costs (VAMOSC).
RAND *RR2510NAVY-2.2*

maintenance, both in the overall account and in individual availabilities. These may be linked in important ways to the overall question of the sustainment phase of LCM.

Figure 2.3, drawn in part from a previous RAND study, shows the annual growth in the overall ship depot maintenance account (account code 1B4B).[2] While the figure depicts all ship depot maintenance, public and private, the relevant component for this study is the green portion labeled "contractor surface," which encompasses all of surface-ship maintenance.[3] In FY 2011–2015, the Navy spent an average of $2.7 billion

---

[2] Martin, McMahon, Kallimani, et al., 2017.

[3] Note that this budget data, while generally comparable to the data in VAMOSC and addressing the same issue, is not the same. The VAMOSC data depicts what was executed while the budget data indicates what was appropriated. However, both show growth.

Figure 2.3
Annual Ship Depot Maintenance Account Expenditures by Ship Class, Civilian, and Contractor, FY 2006–2015

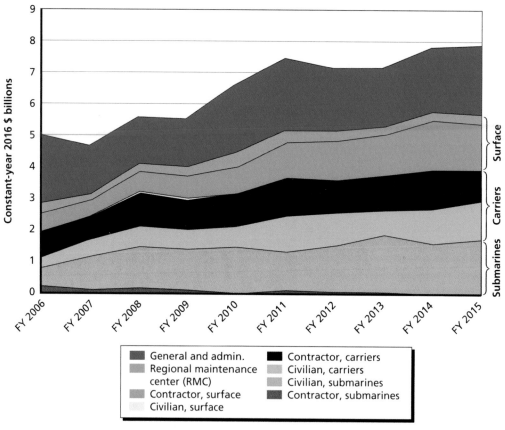

SOURCE: Martin, McMahon, Kallimani, et al., 2017.
NOTE: RMCs provide civilian and military oversight for surface ship depot and intermediate-level maintenance. "General and admin." includes NAVSEA headquarters, other system commands, supply support, transportation, and facilities.
RAND RR2510NAVY-2.3

on private contractor maintenance. From FY 2006 to FY 2010, the figure is approximately $2 billion, although fleet size did not appreciably vary during that period. The report referenced below established that the account growth was largely due to predictable factors in ship maintenance scheduling, in particular the arrival of the DDG-51 and dock landing ship (LSD)–41/49 classes at midlife, at which point both classes underwent large midlife availabilities. These are not unexpected consequences of life-cycle decisions made either in new construction or in sustainment.

However, within the availabilities conducted on surface ships during this period, there was a high incidence of growth (work added to existing jobs due to conditions encountered while performing the job) and new work (work not programmed for the availability but added due to conditions found; new growth is growth within previously unplanned shops). Figure 2.4, also drawn from the RAND study on accounting for growth in ship maintenance accounts, shows the 2003–2015 breakdown of the costs of maintenance by line item for all surface ships. Note that ship's work line-item number (SWLIN) data are from the Navy Maintenance Database (NMD), rather than from the report of overall operations and sustainment cost, which is drawn from the VAMOSC database or from the budget reports drawn from budget exhibits.[4] This breakdown at the equipment category level shows the baseline original work planned for ships within availabilities for this time period. These data reveal that in nearly every category of maintenance organized by SWLIN, there has been a substantial component of growth and variants of new work above the baseline figure. That is, a signifi-

Figure 2.4
SWLIN Growth in Private-Sector Availabilities, FY 2003–2015

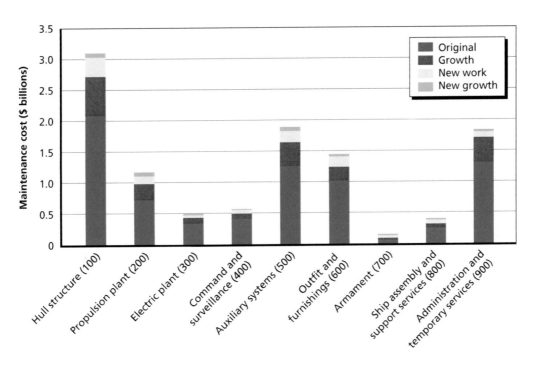

SOURCE: Martin et al., 2017a.
RAND *RR2510NAVY-2.4*

---

[4] Because these databases are kept by different organizations for different purposes, the timelines are different. In the case of NMD, we used all the data that could be recovered, and did not have access to more recent data.

cant proportion of work being done on ships in a depot maintenance period (called "availability") was not planned to be completed and had to be added to after the availabilities started. The portions of the bar marked "original" are what would be expected based on class maintenance plans (CMPs) and technical foundation papers (TFPs), the authoritative technical guides for actions required to allow ships to reach expected service life. The other portions reflect unexpected conditions.

Figure 2.5, also drawn from RAND's report on cost growth in the ship depot maintenance, shows the distribution of overall growth within availabilities by age and by SWLIN. Note that this is growth above the programmed baseline, not the planned work expected from TFPs or CMPs. The high peaks in the graph do correlate with periods when ships expect large baseline availabilities, such as midlife, with the implication that big availabilities experience high levels of growth and new work. There are several possible reasons for this. First, virtually every availability done in private yards experienced significant growth levels, and it is reasonable to assume that larger

Figure 2.5
Cost Growth by Ship's Work Line Item Number and by Ship Age

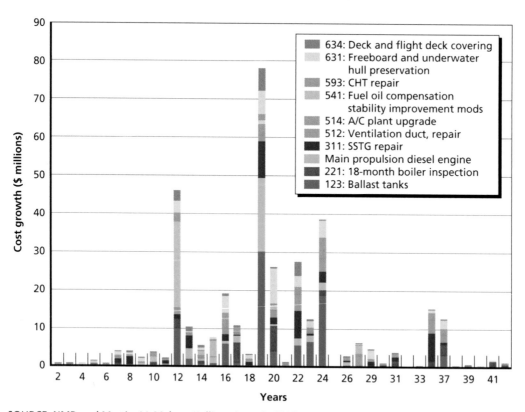

SOURCE: NMD and Martin, McMahon, Kallimani, et al., 2017.
RAND RR2510NAVY-2.5

availabilities might experience proportionally larger growth.[5] Second, more time in a shipyard creates more opportunity to open and inspect systems and find previously undiscovered conditions. Third, ship's force or planners, being aware that a large availability was about to occur, might not pay the same attention to equipment or systems expected to be repaired during availability. In some cases, this could make the problem worse and thus induce growth in the work package.

For the purposes of assessing LCM, the most important issues appear to be in the type of equipment most likely to experience growth and/or new work. Growth and new work are unplanned and thus imply that particular pieces of equipment did not last as long or degraded faster than was expected. As Figure 2.5 shows, this added hundreds of millions of dollars in unanticipated cost to availabilities in execution. At the ship age with the greatest cost increase within availabilities (12, 19, and 24 years), the SWLINs associated with repair of corrosion (123, 512, 541, 593, 631, and 634) account for up to 80 percent of increases. However, there are other significant drivers (diesel engine repair, for example), and these bear detailed analysis for a determination of root cause. By contrast, the root causes for corrosion are relatively simple: coatings do not last as long as expected or preservation was not completed in a timely way. The first relates to items delivered with the ship; the second relates to things needed to be completed as part of sustainment. The data strongly suggest that the Navy is not doing an effective job of managing this process and, as a result, is adding significantly to its sustainment bills in ways it did not expect.

The Impact of Deferral

Corrosion prevention is a straightforward process, but it requires nearly continuous effort. When deferred, the problem gets larger and larger. This is also the case with most other shipboard equipment. While some kinds of equipment may benefit from a "condition-based approach" in which deferral is not necessarily an unwise option, for most equipment, not carrying out maintenance when scheduled might lower short-term costs but exacts a premium when the maintenance is actually done.

The Navy has in fact made a practice of deferral. Figure 2.6 illustrates the difference between maintenance that ought to have been done on ships in the DDG-51 class and the maintenance actually performed. Much of the attempt to recover this maintenance was done in midlife availabilities and that may account for some of the bulge we see in the overall operations and sustainment budget. This is not to suggest that the additional expenditure in 2012 retired the maintenance backlog. Figure 2.6 represents all the maintenance required by TFP relative to all the maintenance performed on all the DDGs in the Navy; this particular figure does not tell us whether some ships were maintained at the expense of others. However, we know from the picture of actual availabilities performed that nearly all of them grew, which at least suggests

---

5   Martin, McMahon, Kallimani, et al., 2017, p. 21.

Figure 2.6
DDG-51 Notional Versus Accomplished Maintenance

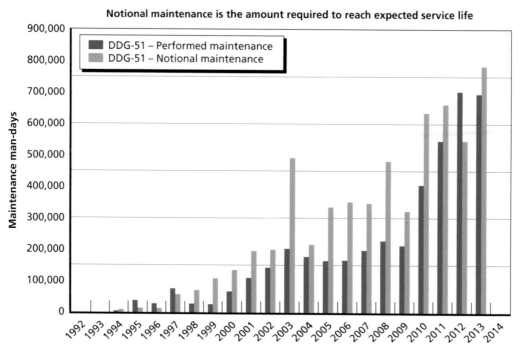

SOURCES: VAMOSC for actual and DDG-51 Technical Foundation Paper for notional.
RAND *RR2510NAVY-2.6*

that the deferral of maintenance relative to TFP is showing up in individual DDG-51 availabilities.

While the history of the DDG-51 class is still being written, the maintenance record of the Ticonderoga class (Guided Missile Cruiser [CG]–47) suggests the effect of deferral on the overall maintenance budget. Figure 2.7 shows that from 2004 to 2011, CG-47 maintenance fell far behind the notional maintenance required to reach expected service life. This disparity was likely due to questions about the class's future and the expectation that these ships were going to be decommissioned in the short term. When it became clear that Congress was not going to allow this action, performed maintenance began to exceed notional maintenance by a significant amount.[6]

Indeed, to demonstrate the effect of deferral on sustainment budgets, Figure 2.8 shows the impact of deferring without attempting to retire any of the deferred maintenance for ten years. While this seems extreme, it is in fact not far from the example we

---

[6]   Bradley Martin, Michael E. McMahon, Jessie Riposo, James G. Kallimani, Angelena Bohman, Alyssa Ramos, and Abby Schendt, *A Strategic Assessment of the Future of U.S. Navy Ship Maintenance: Challenges and Opportunities*, Santa Monica, Calif.: RAND Corporation, RR-1951-NAVY, 2017.

Figure 2.7
CG-47 Notional Versus Accomplished Maintenance

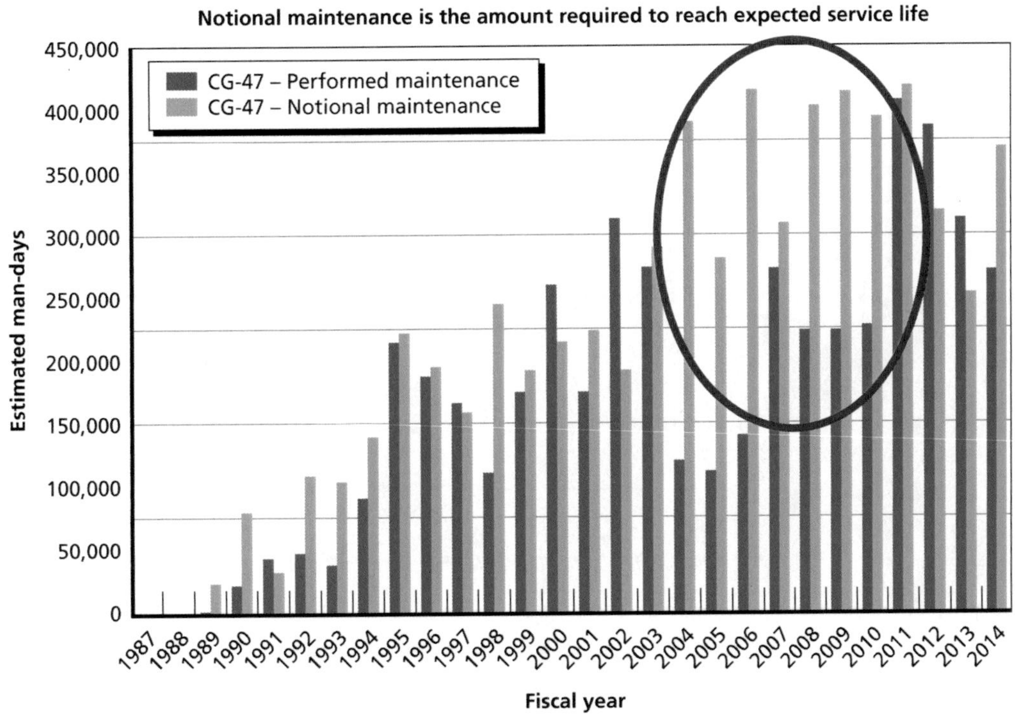

Notional maintenance is the amount required to reach expected service life

SOURCES: VAMOSC for actual and CG-47 Class Maintenance Plan for notional.
RAND RR2510NAVY-2.7

just gave for the CG-47 class. If imposed on the entire Navy, the amount of deferred maintenance would nearly equal the scheduled maintenance for the entire year in 2028. This in effect doubles the out-year demand for maintenance, a result clearly inconsistent with budget planning and satisfactory material condition.

Decisions to defer maintenance reside within the Navy and generally result from a combination of scheduling and funding considerations. Some short-term considerations can force, or at least motivate, paying more in long-term costs for a short-term benefit. While this approach exerts pressure on sustainment costs, it is one over which the Navy LCM enterprise has considerable direct control.

The TFP delineates maintenance requirements that must be done to achieve a ship's expected service life (ESL). Deferring life-cycle maintenance increases costs, work complexity, and time needed to do the work. Implications of the practice of deferring maintenance and then performing "catch-up" of deferred maintenance include excess loading at private yards, delays in delivering ships, and failure of ships to achieve their ESL. Moreover, a domino effect is created to perform this work on all ships due to capacity limits at private yards. As ships are delayed coming out of a depot (due to

Figure 2.8
Effect of Not Retiring Deferred Maintenance

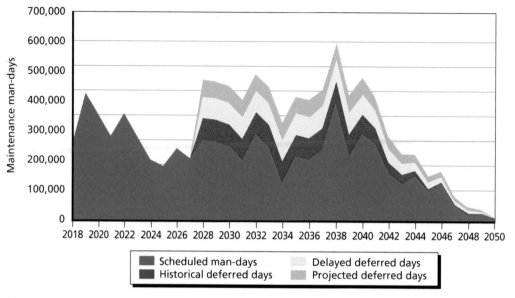

SOURCE: OPNAVNOTE 4700, Class Maintenance Plans.
RAND *RR2510NAVY-2.8*

excess maintenance demands or other reasons), the next ship is delayed in entering the depot. It is challenging for the Navy to catch up with deferred maintenance without an increase in capacity in private yards. The Navy should confirm that TFP requirements are accurate and resource them accordingly, or face the consequences.

Troubled Systems

Depot maintenance data contain a significant level of detail across a number of ships over a number of years. However, they generally do not provide a systems-level view of troubled equipment and, indeed, might not account for equipment that was not repaired specifically in an availability. Some equipment is repaired outside of formal availabilities, sometimes with just the replacement of parts.

The Naval Surface Warfare Center-Corona (NSWC-C) does in fact collect systems-level data and provides this information to Team Ships and other stakeholders in the LCM process. One product of this is a scored "troubled-systems" list, which is based on material casualty report (CASREP) data. These factors include the parts cost, the numbers of failures, the criticality of the systems as measured by effect on operational availability (Ao) where NSWC-C has defined Ao as what the system's availability would have been had the system been fully operational. Table 2.1 shows excerpts of the data contained in the troubled-systems spreadsheet, in this case parts

Table 2.1
Example Data from Troubled-Systems Report

System	CASREP # (Baseline)	CASREP # (FY16)	MRDB Parts Cost (Baseline)	MRDB Parts Cost (FY16)	Ao (Baseline)	Ao (FY16)
Auxiliary seawater	176	156	$10,741,600.00	$12,320,455.00	0.71	0.84
Air conditioning plant	259	208	$6,212,764.00	$9,899,442.00	0.98	0.99
Firemain	190	230	$6,277,951.00	$7,397,906.00	0.98	0.99
Machinery and engineering control	234	223	$5,199,227.00	$5,930,233.00	0.79	0.94
Collection, holding, and transfer	241	230	$9,327,206.00	$10,319,762.00	0.83	0.91
Low pressure air system	252	113	$5,734,171.00	$5,080,285.00	0.98	0.99
Fuel oil	297	281	$9,277,036.00	$11,173,548.00	0.73	0.84
High pressure air system	240	219	$6,619,379.00	$5,229,242.00	0.98	0.99
60 Hz power distribution	79	139	$7,449,748.00	$7,739,027.00	0.94	0.98
Lube oil	214	145	$4,674,079.00	$4,208,624.00	0.86	0.9
Combustion air	53	57	$6,395,682.00	$6,549,860.00	0.99	0.99
Ventilation	331	298	$5,323,063.00	$5,173,519.00	0.65	0.76
Main drainage system	181	204	$2,994,127.00	$2,289,595.00	0.84	0.93
Refrigeration system	139	185	$2,060,699.00	$1,759,470.00	0.89	0.9
Controllable pitch propeller	167	183	$7,828,467.00	$7,423,397.00	0.95	0.96
Chilled water	20	51	$2,120,069.00	$2,542,596.00	0.97	0.98

SOURCE: NAVSEA 21.

cost and Ao for the DDG-51 class. The data reflect a three-year (FY13–FY16) baseline and then a single year (FY16) comparison to the baseline.

These data are then put into an algorithm that weights the different aspects using a priority provided by Team Ships and then results in a top-20 list that enables stakeholders to focus on improving particular systems. The "final rollup rank" indicates a system's overall ranking in the priority list. The example in Table 2.2 is for DDG-51 hull, mechanical, and electrical systems in 2016.

Table 2.2
SEA21 Top 20 Trouble DDG-51 Hull, Mechanical,
and Electrical (HM&E) Systems

Final Rollup Rank	System Name
1	Auxiliary sea water system (AWS)
2	Air conditioning plant
3	Firemain
4	Gas turbine generators
5	Engineering Control System (ECS)/Machinery Control System (MCS)
6	Collection, Holding, and Transfer system (CHT/VCHT)
7	Low pressure air system
8	Propulsion control system
9	Watertight closures
10	Fuel oil system
11	High pressure air system
12	60hz power distribution
13	Lube oil system
14	Combustion air system
15	Ventilation
16	Main/secondary drainage
17	Refrigeration
18	Hull/structure
19	Controllable pitch propeller system
20	Chilled water distribution system

SOURCE: SEA21.

NSWC-C also publishes a summary description of the systems, which explains the features causing them to be troubled. This description is detailed enough to identify in which ships and in what particular system subcomponents troubles were occurring and is restricted as For Official Use Only, so we will not cite results here.

It is important to note that the troubled-systems report and all the reports that drive the algorithm, while generally accurate and detailed, cover only a very limited time frame, comparing one year to a three-year baseline. The troubled-systems list is thus a snapshot driving a short-term work focus, not a long-range planning tool. It is not possible to tell from the fact that auxiliary sea water (ASW) systems are on the troubled-systems list whether this is because of something unusual or simply because a significant number are failing after reaching or exceeding their design life. The maintenance data described earlier and derived from the NMD relate to work items that were not planned (i.e., were encountered unexpectedly). The troubled-systems list does not address this kind of work, but focuses instead on areas where attention is currently needed. This information is helpful, but it is not a guide for improving the life-cycle performance of the equipment.

Troubled Systems and Operational Availability

A CASREP reflects the fact that a system or equipment failure has created a condition that to some degree limits the functioning of the ship. A category-3 or category-4 CASREP indicates a serious or complete mission area limitation and naturally focuses attention on correction. However, CASREPs do not reflect every equipment failure. If a system is redundant and thus does not create an immediate impact on mission accomplishment, a ship might not send a casualty report at all or it might report it as a lower category CASREP. So even if equipment is failing at a higher than expected rate, it might not become part of the troubled-systems list until enough of it breaks to create a crisis. The use of Ao in this context may just say that the casualty is considered to be an operational issue. While that is certainly understandable, it does not really help in identifying approaches to life-cycle issues. Improved Ao in this context generally just means heroic efforts to keep the equipment in operation and likely not improved LCM to gain more operational availability over time.

Ao is most meaningful for a system relative to an established engineering standard. Equipment may be designed to function for some limited number of years, and failure might not be unexpected across some mean range of time. If equipment is failing earlier than the designed life or staying in service longer than the engineered standard would lead us to expect, that information would be useful in helping to plan upgrades. The troubled-systems approach does not include that information; it just tells us that a particular kind of equipment is having an impact on shipboard operations.

Troubled Systems and Cost

Partly by selection by the algorithm, the troubled-systems list clearly reflects concerns about cost. As illustrated in Figure 2.9, the costliest systems are on the troubled-systems list. However, it is not the case that all the troubled systems cost more per system than all the other nontroubled systems in the database. What is most striking about the parts cost per system repair are the cases in which there are few CASREPs: Where those CASREPs are expensive, they are more likely to attract attention than cases in which there are many CASREPs but the cost per unit is not significant.

This emphasis on individually expensive casualties is understandable, but it may in fact conceal the origins of a large proportion of overall parts cost for the Navy. Some systems not on the troubled-systems list but with more numerous failures might in fact be costing more than a few high-profile failures. Figures 2.10 and 2.11 illustrate this. Using data from the VAMOSC system from FY 2010–FY 2015, Figure 2.10 shows not just the cost per system but the average total cost per year of CASREP components; in

Figure 2.9
Parts' Cost per System Repair of DDG-51 Casualty Reports (CASREPs) in FY 2016

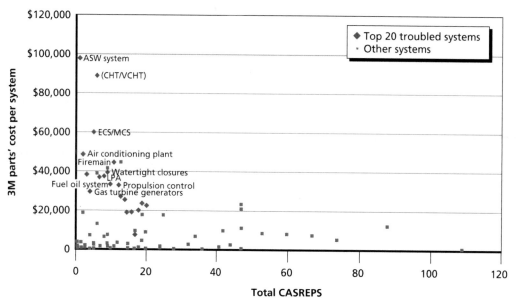

SOURCE: RAND generated from NSWC-C data.
RAND RR2510NAVY-2.9

Figure 2.10
DDG-51 Total Parts Replaced Versus Total Net Cost

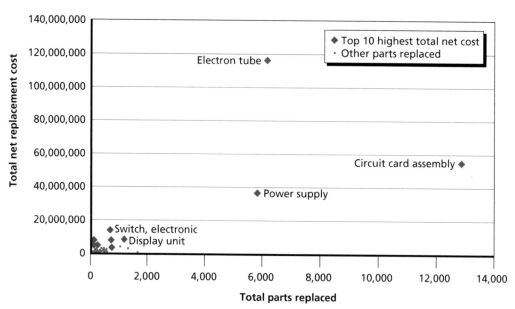

SOURCE: RAND, Navy VAMOSC data.
RAND RR2510NAVY-2.10

Figure 2.11
DDG-51 Total Parts Replaced Versus Net Cost per Replacement

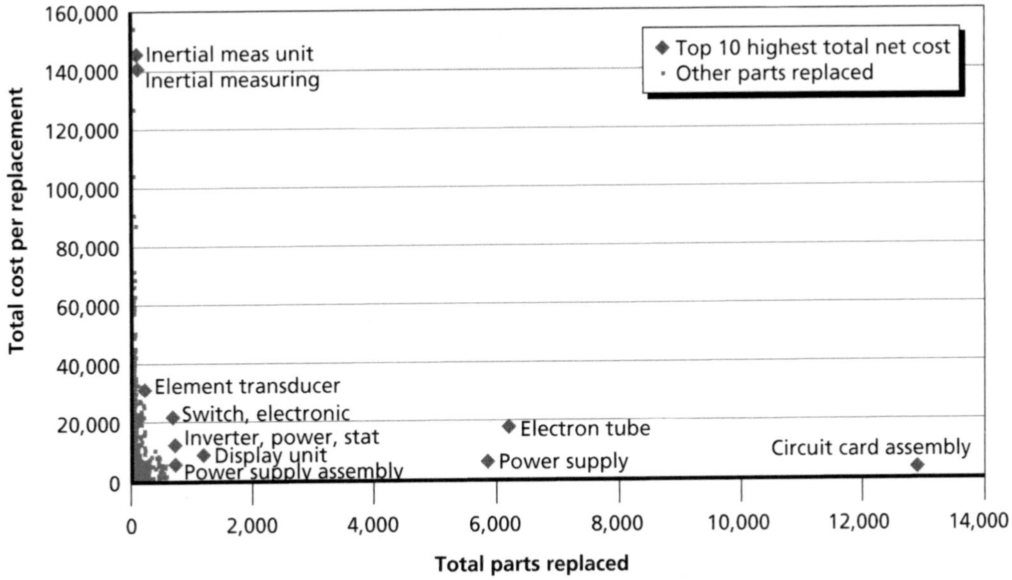

SOURCE: RAND, Navy VAMOSC data.
RAND RR2510NAVY-2.11

other words, it shows the actual cost of replacing a particular component that might have low unit cost but a high rate of failure. By this measure, the parts most needing attention are circuit card assemblies, electron tubes, and power supplies. Individual users might not notice the overall high cost to the Navy, but that cost is large.

Contrast the finding that the parts that may be costliest overall for the Navy may be individually inexpensive but collectively costly with Figure 2.11, which shows the highest parts costs per unit compared with the numbers of parts replaced. This view would push the user toward concentrating on high-visibility failures in which a small number of units cost a large amount. By this metric, several of the most expensive total cost items would likely be lost in the noise of similarly priced components, and two very expensive individual units (inertial measuring units) would appear to be major cost drivers, when in fact they barely breach the surface in Figure 2.10, which shows overall cost to the Navy. An organization that was mostly reacting to obvious inputs might spend considerable time and attention on inertial measurement units and less on electron tubes, power supplies, and circuit card assemblies, which is where the majority of resource expenditure is taking place. The greatest reduction in parts cost would come from lowering the cost or increasing the life span of the most common items, rather than from focusing on less common, high individual cost items.

Differences between databases may, however, limit the applicability and reliability of the finding that some of the most numerous components are collectively the most costly. The costliest parts identified in the data supporting the troubled-systems list may be represented by different national item identification numbers (NIINs) in the VAMOSC database. For example, in the VAMOSC data used in this study, we found 1,543 separate NIINs for DDG-51 parts labeled "circuit card," 260 for "power supplies, and 18 for "electron tubes." The difference in NIINs results from the parts being used in different systems; parts having the same descriptor but accomplishing different tasks (e.g., power supplies producing different voltages, amperages, and phases); multiple manufacturers making interchangeable parts; parts being phased into the system; and parts being phased out of inventory. Connecting the NIINs to recognizable part names was accomplished using a separate database, Naval Supply Systems Command (NAVSUP) One Touch, which required additional permissions to access: a general user of VAMOSC or the troubled-systems list would not necessarily have access to these data. Developing a complete picture of maintenance cost drivers was not possible without consulting multiple data sources.[7]

What Do We Know About the Sustainment Phase of Life-Cycle Management?

The most obvious observation is that several different data sets tell us different things about the LCM of ships in service. They are generally not connected to one another, and the result is an ad hoc series of processes to deal with different parts of the problem. But a few things stand out:

- Depot maintenance data sets do provide a longitudinal view, and they show that maintenance costs have steadily escalated and that a significant proportion of this cost is associated with unplanned work. Although this is probably true across many categories of equipment, it is most obvious from the data on hull structures and tanks, places highly susceptible to corrosion, which is likely to go grow in severity the longer work is deferred.

- Troubled-systems approaches can provide rich information on material readiness issues currently preoccupying the Navy, but the current data sets are limited in time on their horizon, reflect a preoccupation with getting ships fixed and not with understanding why they are broken, and are not effective in tracking long-term parts and repair issues. The major observation concerning the troubled-systems approach is that it builds into its algorithms priorities that focus on

---

[7]   Navy VAMOSC can provide access to Open Architecture Retrieval Systems (OARS) data, which connects to Commercial and Government Entity (CAGE) code, as well as to part numbers and NIINs listed with jobs in shipboard Current Ships Maintenance Project. This is a multistep process requiring access to several databases and providing narrowly scoped answers. This functionality might help, but will not overcome, the need to consult multiple data sources and systems.

expensive but not frequently fixed parts, as well as parts that have degraded to the point that prevention is no longer an option.

- The troubled-systems process focuses attention on high cost-per-unit items, not on units that, though individually inexpensive, might be those most affecting life-cycle costs due to their high failure rate. While it may be that the NAVSUP or other logistics agencies are attempting to identify these types of parts, we did not find any evidence that the ship LCM enterprise is attempting to address this issue. In any case, the troubled-systems approach would not yield much information about this issue.

The major conclusion from a review of the evidence on sustainment is that there is a clear effort on the part of the Navy to deal with material issues that arise from increasing depot maintenance funding and troubled systems. There is, however, an apparent lack of focus on the longer-term issues impacting LCM costs. We examine the organizational and incentive structures that might contribute to this condition in the following chapters. However, even as the study team assessed the state of sustainment readiness, we found a multiplicity of databases that yield a variety of answers on different aspects on material issues of ships in service. This is manifestly an issue, and we will return to it. Not having a coherent and integrated data set from the outset makes it impossible to readily assess the impact that different decisions might be having on different parts of the sustainment problem.

Sustainment Trends and the Relationship with New Construction

We discuss the process for developing the concept and requirements for new ships in detail in the next chapter. However, the results shown in this chapter do suggest that the feedback process shown in Figure 1.1 is not generally occurring, at least in terms of flow from ships in service back to new construction. This conclusion is based on three primary observations:

- The escalation in ship maintenance cost continued unabated through several years, and there does not appear to be a plan in place to address the underlying reason for issues such as corrosion during new construction. Some of these issues could be corrected during construction by different designs and different coatings. There does appear to be an effort to use better coatings.[8] But the latest ship class developed by the Navy, the Littoral Combat Ship (LCS), is beset by propulsion system design and hull structure corrosion issues, which, in turn, suggests that the sustainment needs of the fleet are not being closely attended to.[9]

---

8   Tim Pennington, "Navy Is Full Steam Ahead on Powder Coating," *PF: Products Finishing*, February 1, 2012.

9   Sydney J. Freedberg, "LCS Troubles May Stem from Double Engine," *Breaking Defense*, September 7, 2016.

- The consolidated view of ship troubled systems has a very limited three-year time horizon, oriented mostly toward identifying and fixing immediate problems. Without significant additional analysis, it would not even be appropriate to use this list as the basis for recommendations for ships in new construction.
- Similarly, the systemic attention to equipment that costs a lot but might not break very frequently might detract from examination of common, less expensive parts that do break often and ultimately become more costly to repair. Shifting to focus to these parts would in fact more effectively reduce long-term costs.

We now examine in more detail the types of decisions and processes used in new construction.

CHAPTER THREE

New Ship Construction

Our study focuses on improving connections between processes in different phases of the ship's life cycle. The previous chapters identified several challenges in sustainment. In this chapter, we explore the new ship construction process and identify stages of shipbuilding where better process and data could improve the life-cycle sustainment outcomes. We also discuss challenges associated with translating such recommendations into new construction processes.

The New Ship Construction Process

Ship construction programs, like all Major Defense Acquisition Programs, go through a process of requirements identification, material solution identification, and capability specification before construction even begins. Once this process is completed, Navy new construction shipbuilding processes can then be divided into five stages: design, construction, test, delivery, and postdelivery. After postdelivery the Navy provides the ship to the fleet. Figure 3.1 shows the phases and their interrelationships.

Design Phase

The design phase has several stages. These include the engineering design stage, which consists of preliminary design and contract design, and the production design stage, which consists of detail design and construction.[1] Before engineering design stage, there is an exploratory/concept design process, during which technology is assessed and alternatives are analyzed in order to decide which material solution to pursue for the result of the acquisition life cycle.[2] During the design phase, hull systems, mission systems, propulsion/power/machinery (PPM), human systems, survivability, and design integration and management areas are all examined.[3] Trade-offs and iterations occur

---

[1] Naval Sea Systems Command, *The Navy Ship Design Process*, Naval Surface Warfare Center-Carderock, Bethesda, Md., January 6, 2012.

[2] NAVSEA, 2012.

[3] NAVSEA, 2012.

Figure 3.1
The New Ship Construction Process

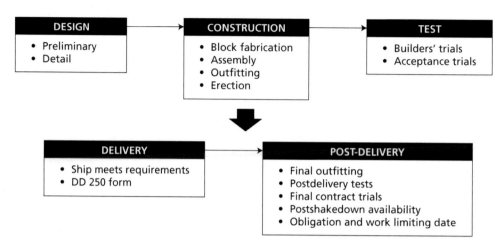

SOURCE: Adapted from U.S. Government Accountability Office, *Navy Shipbuilding: Opportunities Exist to Improve Practices Affecting Quality*, Washington, D.C.: GAO-14-122, November 19, 2013, pp. 9–11.
RAND *RR2510NAVY-3.1*

between the design areas until a baseline has been reached.[4] By the end of the design phase, 80 percent of the future life-cycle costs, both known and unknown, are established to the point that the organizations responsible for sustainment will be unable to affect them significantly once the ship is delivered.[5]

The goals of the design phase are to take the ship specifications developed through the requirements process and translate those into detailed ship designs. The ship design manager (SDM) is responsible for design and integration efforts with assistance from the Supervisor of Shipbuilding Conversion and Repair (SUPSHIP) Chief Engineer (CHENG) as outlined in NAVSEAINST 5400.95F. The two organizations work together to develop drawings, models, schematics, and lofting packages and to identify long lead-time material.[6] The technical authority pyramid is shown in Figure 3.2.

During this period, many CSE decisions are made. Since the design of the ship will dictate the quantity, location, and use of various common systems, feedback during this stage about CSE already in the fleet would be most useful (e.g., items that are failing more often than expected, items that were used in a novel way/location and are succeeding or not succeeding, and so on). Since 80 percent of the life-cycle costs

---

[4] NAVSEA, 2012.

[5] U.S. Government Accountability Office, "Navy Actions Needed to Optimize Ship Crew Size and Reduce Total Ownership Costs," Washington, D.C.: GA0 03-20, June 2003.

[6] Naval Sea Systems Command, Supervisor of Shipbuilding, "Engineering and Technical Oversight," *SUPSHIP Operations Manual (SOM)*, Chapter 8, S0300-B2-MAN-010, Bethesda, Md., August 28, 2015.

Figure 3.2
Technical Authority Pyramid

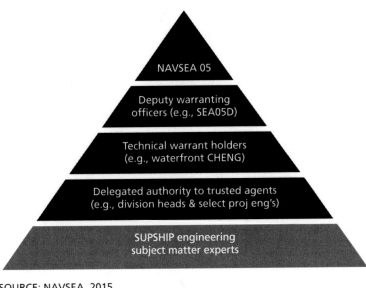

NAVSEA 05

Deputy warranting
officers (e.g., SEA05D)

Technical warrant holders
(e.g., waterfront CHENG)

Delegated authority to trusted agents
(e.g., division heads & select proj eng's)

SUPSHIP engineering
subject matter experts

SOURCE: NAVSEA, 2015.
RAND RR2510NAVY-3.2

are locked in at the end of this phase,[7] life-cycle cost risks associated with equipment unique to a class or unproven CSEs should be assessed at this time too. Where the cost risk exceeds reasonable bounds, design changes should be made.

Also, during this phase, the shipbuilder develops an integrated master schedule and a configuration management plan, which provide baselines for the design. Any modifications to the design must go through the program manager. After design is baselined, the ship is ready to go to construction.

Construction Phase
During the design phase, the ship-class program manager approves contracts that dictate which items the shipbuilder is responsible for procuring/building and which items will be contractor-furnished or government-furnished. A multitude of organizations in addition to the equipment manufacturers, shipbuilders, and program managers are involved in the ship construction process (see Figure 3.3 below), and there are military, commercial, and regulatory standards (e.g., American Bureau of Shipping, Department of Energy, etc.) that must be adhered to during construction.

---

7 U.S. Government Accountability Office, 2003.

Figure 3.3
Organizations and Individuals Involved in Ship Construction

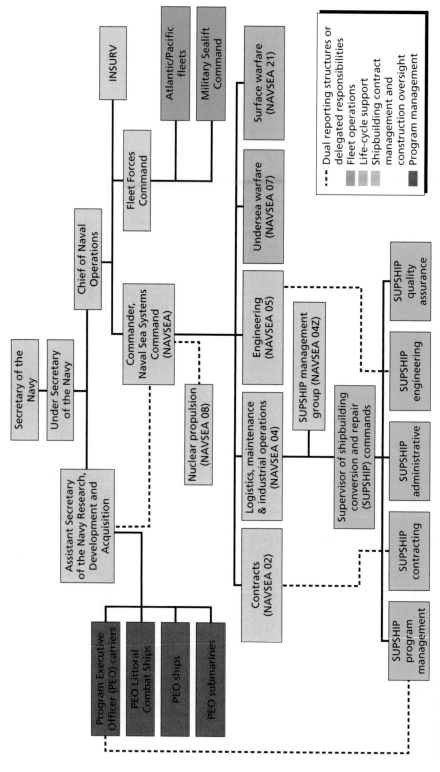

SOURCE: U.S. Government Accountability Office, *Navy Shipbuilding: Opportunities Exist to Improve Practices Affecting Quality*, Washington, D.C.: GAO-14-122, November 19, 2013.

RAND *RR2510NAVY-3.3*

Given the breadth of organizations involved in the construction process, the Navy has attempted several initiatives to improve oversight during the construction process. According to the Government Accountability Office (GAO),

> In 2009, the Navy Organization that oversees ship construction launched the Back to Basics initiative to improve Navy oversight of ship construction. However, a key output of the initiative promoting consistent and adequate quality requirements in Navy contracts has yet to be implemented.[8]

An example of the impact of poor consistency and quality management in Navy contracting is the practice of contractors obtaining and storing materials for years prior to installation. Often those materials are obsolete before they are installed in new construction ships. It is during this phase that consolidated obsolescence information, updated life-cycle cost estimates for component parts, and additional feedback from the fleet for currently used parts would be particularly useful. Absent this information, shipbuilders and contractors may purchase and store approved parts for many years; when the equipment is finally installed on the ship, it is already obsolete or no longer meets Navy standards.

Test Phase

After ship construction is complete, system testing and sea trials begin. Tests run all the way through postdelivery (postdelivery tests, final trials, and postshakedown availability). The sea trials (builders' trials and acceptance trials)[9] are performed to compare the ship's performance with the contract specifications. The ship buyer (program manager) and shipbuilder are present for the builder's trials, while the Board of Inspection and Survey (INSURV) conducts the acceptance trials. The Joint Test Group (JTG) consists of persons assigned by their parent organizations to make required local approval actions for a test program.[10] JTG members include the shipbuilder, ship's force, SUP-SHIP, and other organizations with ongoing work and testing responsibilities.

Test activities may include the following:[11]

- review and approval of test procedures
- review and approval of test problem reports
- review and approval of test reports.

---

[8] U.S. Government Accountability Office, 2013.

[9] "Builder's trials test the vessel's propulsion, communications, navigation and mission systems, as well as all related support systems. Following the successful completion of builder's trials, acceptance trials are conducted by the Navy's Board of Inspection and Survey (INSURV)" (U.S. Government Accountability Office, 2013, p. i).

[10] NAVSEA, 2015.

[11] NAVSEA, 2015.

Test events may include:[12]

- flood dock
- undock (or launch)
- initial criticality
- engine light-off
- dock trials
- fast cruise.

Data from these tests provide some of the first operational test data for the ships. These data can be used to modify future ships in the same class and future CSE processes across all classes. However, despite the obvious value this data might have, no database exists that stores information on how individual systems and pieces of equipment fared during testing.

Delivery

After a ship has completed the testing period, the "buyer takes custody and assumes ownership of the vessel."[13] This is known as "ship delivery/acceptance." Form Defense Form (DD) 250 represents the handover from the shipbuilder to the Navy and documents any remaining work from the construction or testing period as well as the responsible party. The Chief of Naval Operations (CNO) decides when to transfer the ship from the builder to the program manager. After the CNO approves delivery, from this point forward the program manager is responsible for the ship until it is handed off officially to the fleet. Figure 3.4 shows the roles of the Navy organizations involved in ship delivery.

Postdelivery

After the ship is delivered, but before it is handed over to the fleet, the Navy must finish any work that was not completed during construction, perform modernizations and upgrades to replace existing equipment, and undertake new work to add or change equipment in the ship's baseline. Figure 3.5 below illustrates the activities in the postdelivery period.

Because multiple work efforts are usually being undertaken simultaneously during postdelivery, multiple types of appropriations fund these activities. Table 3.1 lists the types of appropriations used during postdelivery.

---

[12] NAVSEA, 2015.

[13] U.S. Government Accountability Office, 2013.

Figure 3.4
Navy Organizations Involved in Ship Delivery

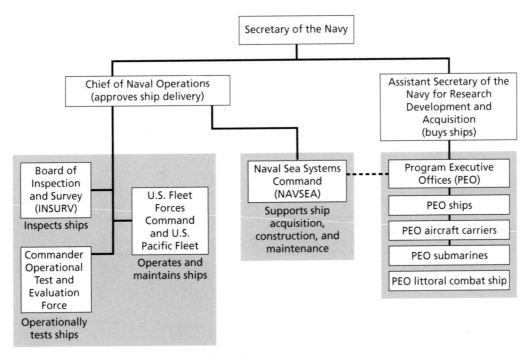

SOURCE: Government Accountability Office, 2013.
RAND RR2510NAVY-3.4

Figure 3.5
Postdelivery Period Activities

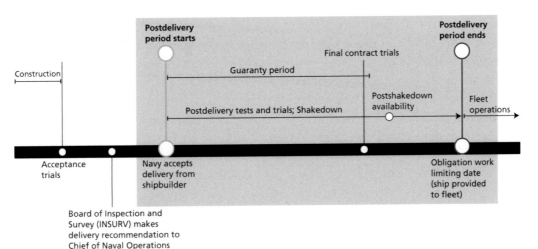

SOURCE: U.S. Government Accountability Office, "Navy Shipbuilding: Policy Changes Needed to Improve the Post-Delivery Process and Ship Quality," Report to Congressional Committee on Armed Services, U.S. Senate, Washington, D.C.: GAO-17-418, July 2017.
RAND RR2510NAVY-3.5

Table 3.1
Postdelivery Appropriations and Activities

Funding Category	Description
Shipbuilding and Conversion, Navy (SCN)	This is the main account used prior to the obligation work limiting date (OWLD), and is used for, among other things, ship construction and system installation (procurement). SCN funding is also used for • outfitting, which involves acquiring on-board repair parts, such as valves, and • postdelivery activities such as correcting deficiencies and conducting tests and trials. What is referred to as a ship's "end cost" generally includes funds used for ship construction, deferred work, and change orders.
Operation and Maintenance (O&M) Navy	This is the main account used after OWLD and is used for the day-to-day costs of operating naval forces. During the postdelivery period, operations and maintenance funding is used to support the ship's crew and pays for consumables such as fuel and fleet-responsible maintenance.
Research, Development, Test and Evaluation, Navy	This account is used for research, development, test, and evaluation efforts performed by contractors and government installations to develop equipment or purchase materials, weapons, or computer application software. These efforts may include purchases of services (such as engineering), which occur throughout the shipbuilding process, including during the postdelivery period.
Other Procurement, Navy (OPN)	This account finances the procurement, production, and modernization of equipment not otherwise provided for. During postdelivery, OPN is generally used to fund Chief of Naval Operations–sponsored upgrades.

SOURCE: U.S. Government Accountability Office, 2017.

The Link Between New Construction and Sustainment

We have outlined the process currently used for ship construction decisions. However, we have not considered how this is intended to relate to the sustainment process we discussed earlier. Selecting known but less-than-optimal CSE can result in obsolescence costs, early modernization/replacement costs, and frequent repairs. Selecting unknown equipment can result in a loss of performance ability and pose the same risks as the known items listed above. As discussed in Chapter Two, the Navy has multiple processes to correct deficiencies once they are identified and spends a steadily escalating amount to support these processes.

Our research showed that the Navy does not have similarly robust processes for linking decisions in new construction with potential life-cycle impact later. The description of construction processes provided above indicates that the best times for influencing choices of CSE are during design and construction phases, when decisions on equipment performance and configuration are being made. There is, however, no specific process or provision for ensuring that sustainment decisions are considered in these phases.

Ships and major equipment on ships are delivered according to a set of requirements that are specified at the highest level with key performance parameters (KPPs) contained in capability development documents (CDDs) and then at successively more detailed specification levels. Generally, these KPPs are assessed in the test phase. However, while KPPs for system reliability are generally assigned, they are not the types of values that can be readily assessed in testing. Some combination of modeling and long-term assessment is required, and, unfortunately, long-term assessment takes the ship and its equipment long past the phase in which new decisions could be made in new construction. Thus, equipment testing is also not likely to ensure full consideration of reliability or sustainability early in the ship's life cycle. In fact, we describe several examples in the next section that suggests this link is seriously flawed.

The following examples are not presented as an exhaustive list or a set of detailed case studies. All these cases have been studied in some detail, and we are not attempting to address all the acquisition problems that might have arisen as these ships were designed, built, and delivered to the fleet. We do, however, believe these represent some salient points about the disconnects between phases of the life cycle. It is important to bear in mind that there are very limited numbers of surface ships undergoing construction at any particular time, and the issues we identify are generally class issues rather than specific ship issues. There is no controversy that the issues discussed below are major ones that affected LCM. The question is how these affected subsequent LCM phases after delivery, and whether there was evidence of feedback in the system to correct them once identified.

New Construction Issues Affecting Life-Cycle Management

The process outlined above arose from a series of reforms in Navy shipbuilding, whose goals were to reduce the total life-cycle cost of ships by offering providers better incentive structures and appropriate levels of oversight. These reforms should also ensure that newly constructed ships are delivered with minimal defects and quality issues. However, a recent GAO report found that these goals are not being met.[14] Even with all of this work done during the postdelivery period, ships are still delivered to the fleet with uncorrected deficiencies.

These new construction quality issues do not have clear solutions at the shipbuilder, program manager, fleet, or resource manager level. They add expense that might not be captured in initial ship construction, might delay the delivery of the ship to the fleet (and thus increase the demand for maintenance-intensive legacy platforms), or might impose costs for repair that could have been avoided by different new construction decisions.

---

[14] U.S. Government Accountability Office, 2017.

Figure 3.6
Number of Problems or Defects by Phase

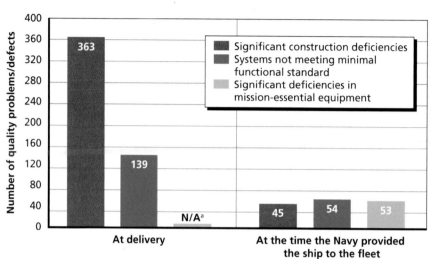

SOURCE: U.S. Government Accountability Office, 2017.
ª This information is not evaluated at delivery.
RAND RR2510NAVY-3.6

Figure 3.6 shows the prevalence of these problems. The majority of problems are identified before delivery, and so the major effect is generally on the expense of ship construction or the delay in getting the new ship to the fleet, rather than directly on sustainment. However, this is not always the case, and defects can carry over into the period when the ship is supposed to be performing operational missions for the Navy.

Four Categories of New Ship Construction Issues

Four specific issues were selected for their frequency, their potential cost, and their effect on subsequent sustainment. In each case, we give examples illustrating details of the issues identified. However, our purpose is not to affix blame for earlier decisions or describe them in great detail; that has been done elsewhere by GAO and the Navy itself. Rather, what we want to understand is where in the process Navy organizations might be able to intervene and how this intervention might be facilitated. We thus need to describe the circumstances well enough to allow an organizational analysis. We divide these issues into four broad categories:

1. construction problems/defects at delivery (equipment is defective because of shipyard/contractor construction issues)
2. ship delivered with equipment already over budget (equipment selected is too expensive)

3. ship delivered with equipment that is does not work because of technological optimism (equipment selected is not technologically mature by ship delivery)
4. ship delivered with equipment that does not work because of planning/design errors (equipment selected is nonfunctioning because of planning/design errors).

Construction Defects

The LPD17 class provides several examples of a construction defect that carried over into sustainment. LPDs are amphibious transport dock ships that "embark, transport and land elements of a landing force for a variety of expeditionary warfare missions."[15] The first ship of the class, the USS *San Antonio*, suffered failures, particularly in main propulsion, was delayed in delivery, and then had to be entered into an availability after delivery to correct multiple issues in its main propulsion diesel engines.[16] This unexpected delay resulted in changes to deployment schedules, a requirement to keep the LPD4 class long past its expected service life, and diversion of maintenance resources. Despite a relatively clear picture of class deficiencies, LPD21 was delivered with over 6,000 deficiencies (see Figure 3.7). Although the number of deficiencies has lessened on recently delivered ships, the LPD24 was still delivered with over 1,000 deficiencies attributed to the shipbuilder.[17]

Figure 3.7
Deficiencies at Time of Delivery for LPD17 Ship Class

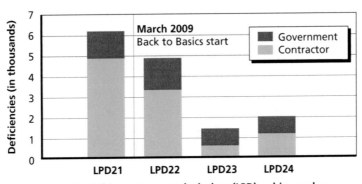

SOURCE: GAO analysis of Navy data.
RAND RR2510NAVY-3.7

---

15 "Amphibious Transport Dock—LPD: The Most Powerful Amphibious Force of All Time," *United States Navy Fact File*, May 2, 2018.

16 Michelle J. Howard, *Command Investigation of Diesel Engine and Related Maintenance and Quality Assurance Issues Aboard USS* San Antonio *(LPD17)*, Norfolk, Va., Department of the Navy, January 15, 2010.

17 U.S. Government Accountability Office, 2013.

These deficiencies ranged from foundation bolts that were not fitted properly and resulted in engine movement during normal operation to using the wrong weld material on exhaust systems and thus causing corrosion and insulation burns.[18] The LPD25 was also delivered with a contractor-furnished electronics system that experienced "widespread performance failures and the Navy has been unable to repair the ship efficiently, including during the post-delivery period and after the ship was provided to the fleet. As a result, the Navy is in the process of looking at incorporating a new system."[19]

This class ought to have been a good place for the feedback model we described in Chapter One to occur. However, this does not appear to have been the case, at least not until the eighth in an eleven-ship class was reached.

Ship Delivered with Equipment Already over Budget

Sometimes ships are delivered with equipment that is considerably more expensive than originally budgeted. This might generally be absorbed as part of overall construction cost, and thus would not necessarily become an issue in later sustainment. However, if the expense to install is too great and cannot be done on time, the ship might simply be delivered without complete capability. Or, alternatively, the ship may be delivered with the capability, but late and with resulting impact to ships required to remain in service for a longer period.

This problem generally does not occur for contracted items, as there is contractual recourse for failing to deliver capability. In the period we examined, this type of cost escalation came primarily from unexpected cost growth or delay in government furnished equipment (GFE). In the case of the Ford-class aircraft carrier, delays and cost increases in two critical GFE subsystems—the electronic magnetic launching system (EMALS) and the advanced arresting gear (AAG)—were the major reason the class continues to be over budget and late.[20] This in turn impacted deployment schedules of ships in service and required ships advanced in service life to remain in service longer.

The cases in new surface-ship construction are not as dramatic, but they do exist. The DDG-1000 was conceived as a land attack destroyer whose major mission is support of expeditionary forces ashore with high volume fires. It uses the "Advanced Gun System," which was intended to deliver a "Long-Range Land Attack Projectile" (LRLAP). Originally, a single LRLAP round was supposed to cost approximately $50,000.

---

[18] U.S. Government Accountability Office, 2013.

[19] U.S. Government Accountability Office, 2017.

[20] Bradley Martin and Michael E. McMahon, *Future Aircraft Carrier Options*, Santa Monica, Calif.: RAND Corporation, RR-2006-NAVY, 2017.

The round instead escalated in cost to approximately $1 million per round, resulting in the Navy's canceling the LRLAP program in 2016.[21] This cost increase was not due to technological development issues or shipbuilder or weapons developer performance. The gun and the round were both performing well in testing. However, the LRLAP program was originally intended to support a 32-ship class, and the dramatic reduction in quantity of the ship class to three drastically escalated the unit cost. None of the efficiencies of quantity order was available, and there was no ability to spread the development cost across the class.

This meant that the cost of providing the individual ships with a full weapons load-out became unduly expensive, and so now the ship is being delivered with a gun system that effectively has the same range as legacy platforms. The Navy will either have to accept the limitation or find some different solutions. In fact, a less costly option for ammunition was available—the Excalibur guided artillery round, which provides roughly half the range of the LRAP—but this would have required modifications to the gun barrel, cooling systems, and automated magazine, which would have involved an unknown but likely significant expense; one estimate of engineering design costs to modify all three ships in the class to accommodate different sizes and projectile types came to $250 million.[22] The last alternative would be to completely replace the gun system, with unknown effect on cost, schedule, and sustainability.

The specific organizational failures in this case appear to be related primarily to the setting and funding of requirements. The LRLAP is an ammunition round, and is thus more expendable than a piece of installed equipment. Without it or similar capability, however, the DDG-1000 is incapable of performing the mission it was designed to do. There might have been choices earlier in the ship's design and construction process that would have allowed a less expensive round with lower development costs. That does not appear to have been part of the discussion, and, as a result, not only is the ship itself extremely expensive, but anything to make the gun system operational will also be extremely expensive.

Nonfunctioning Equipment (Technological Optimism)

This problem occurs when some piece of equipment held to be essential for the operation of a new ship turns out to be infeasible due to difficulties developing or integrating new technology. The LCS classes offer several examples in connection with the mission modules required to enable this ship to perform more than simple presence missions. The mine countermeasures module, for example, suffered multiple delays due to difficulties with the mine-hunting vehicle, mine-hunting sensors, and neutralization

---

21 Kyle Mizokami, "The USS *Zumwalt* Can't Fire Its Guns Because the Ammo Is Too Expensive," *Popular Mechanics*, November 7, 2016.

22 Sam LaGrone, "Raytheon Excalibur Round Set to Replace LRLAP on Zumwalts," *USNI News*, January 1, 2017.

capabilities.[23] Although these technologies were seemingly not extremely advanced or sophisticated, there were difficulties integrating them together and with the platform. Even if the LCS had not encountered other difficulties with production and delivery, the module shortfalls effectively preclude it from carrying out projected war-time missions.

Other examples of unfounded technological optimism have occurred in ship propulsion systems on LCSs, boat-handling and recovery equipment on LCSs, and integrated ship control systems on the DDG-1000. In some cases, the result of technological optimism is less in the loss of needed capabilities and more in the inability to achieve expected savings in sustainment from less maintenance-intensive systems or decreased manpower. It is unclear, for example, in the case of the LCS that a more highly automated ship with a smaller crew has in fact resulted in less demand overall for maintenance personnel.

Nonfunctioning Common Shipboard Equipment (Design/Planning Errors)

Both the LCS and the LPD17 suffered numerous propulsion casualties before and after delivery. Some of these are likely attributable to poor contractor performance and delivery with manufacturing defects. However, part of the problem is clearly a matter of design. LCS-1's machinery plant is cramped and difficult to access. Both LCS variants have complicated gas turbine–diesel plants with poorly performing machinery interfaces.[24] Both are equipped with water-jet propulsors that rapidly break down the underwater paint coatings, accelerate corrosion, and require that LCSs be dry-docked about twice as often as other ships of comparable size. This is not the result of poor performance by a contractor; the design was flawed, and the long-range planning that should accompany delivery of the ship had gaps.

In all the LCS cases, it is important to note that PEO LCS is a separate organization from Team Ships and that the program has been beset by well-publicized failures that do not necessarily apply generally. However, similar process failures were present in the LPD17, to a degree in the DDG-1000, and even in well-established programs for big-deck amphibious ships. Our underlying point is not to search for episodes of failure but to note where the failures occurred and look for remedies. A frigate based on an LCS hull is planned; requirements for future surface combatants are being assessed and prepared. The follow-on to the LSD-41/49 class will be based on the LPD17 hull and propulsion plant. Assessing performance, capturing lessons learned, and looking for process improvements seem particularly important at this time.

---

[23] FY 15 Navy Programs, "Littoral Combat Ship (LCS) and Associated Mission Modules," undated.

[24] Franz-Stefan Gady, "Dropping Like Flies: Third US Navy Littoral Combat Ship Out of Action," *The Diplomat*, August 13, 2016.

CHAPTER FOUR

Improving the Ship Equipment Life-Cycle Process

We have identified a number of issues in both the construction and sustainment phases of ship life cycles. The major theme in each of these is that the Navy is reacting to problems it can identify, but it has difficulty anticipating problems and sometimes seems unable to identify longer-term trends that may actually be driving costs. For example, the data on expense for parts yields the finding that the greatest expense is in low cost-per-unit equipment with a high failure rate. The items receiving attention in the troubled-systems list are individually the most expensive, but that might not be the best place to focus effort. Similarly, the unexpected cost growth in ship availabilities comes principally from a phenomenon that should be wholly expected: corrosion in tanks and on hull structure. The new construction examples show a whole series of decisions that are effectively guaranteed to create sustainment problems.

The decision to follow some particular path was not made in a vacuum, and an understanding of how these conditions came to be and how they might be corrected is the major research question of this study. It does not in general appear that the conditions we have observed are due to technical issues that could not have been anticipated, nor does it appear that the reactive solutions the Navy implemented would have been difficult to conceive before they became an after-the-fact reaction. It is not that the problems are beyond the technical capability of the Navy to resolve; it appears more likely that organizational issues are hampering the ability to carry out a cohesive response.

The Organizational Problem

The Navy shipbuilding and sustainment enterprise is large, and it is common for different imperatives and cultures to develop in large organizations. It is not an inherently bad practice for one organization to be concerned with the design and construction of ships and another to be concerned with all the issues of sustainment. There will also likely be tension between those who establish requirements, those who program and budget resources, and those actually executing either ship construction or sustainment. However, it is clear from the data we have presented that something is dysfunctional in the surface warfare enterprise, and organizational structures and institutionally driven incentive structures appear to be likely contributors. There is no doubt that difficult technical and resource challenges exist in both ship construction and ship sustainment. But when problems arise, solutions and resources are found, so it appears that the effort should be arriving at solutions before rather than after the fact.

Three Focus Areas

We examine three areas where organizational practices are hampering a broader and more systematic approach and will suggest possible solutions. These areas, which we selected based on the patterns observed in the data analysis and research we and others have performed, are the following:

- data reporting and compatibility
- funding and incentive structures for short- and long-term focus
- lack of common command perspective.

Regarding the first, the number of different data sets that different actors in the process use is staggering. As we attempted to look for trends in LCM, we were faced with multiple databases, all of which deal with some portion of the life-cycle process, but are not in any meaningful way interoperable with each other, and they also have differing access restrictions. We also noted that there were many long-term problems, but no obvious attempt to generate long-term solutions. The incentive structures guiding decisions in nearly every aspect of ship LCM do not promote a long-term view. Finally, while many commands are involved in building, maintaining, and operating surface ships, there is not a common commander, and, as a result, little incentive exists to seek an integrated view.

Data Compatibility Issues

Our analysis in Chapter Two showed trends in different readiness areas. However, the striking feature the analysis revealed is that so little of the data sets are common or compatible. VAMOSC, the Navy's program of record source for operations and sustainment data, appears to show something different about the relative cost of systems than does the parts tracking system used by NAVSUP. The CASREP data maintained by NSWC-C provides a snapshot for a limited number of years, but it does not provide much insight into trends. Maintenance data from multiple sources show that hull structure, which contributes most to total life-cycle costs, is markedly absent from the Navy's top troubled-systems report. These findings—and the need to employ multiple data sources to uncover them—suggest a fundamental problem with the current approach to data management: The existence of numerous disparate data sources makes it a challenge for stakeholders to develop a complete picture of cost factors.

Reasons for the existence of disparate data sources are understandable. Numerous realignments over the years intended to address various problems with O&S have produced multiple reports, databases, software, and websites with different proprietors and users. Several of these data sources that are relevant to LCM are described in Table 4.1. Collectively, these sources offer a wealth of useful information for understanding LCM issues, but individually they are stovepiped to their specific target organizations. As a result, different stakeholders in the LCM group see different information.

Table 4.1
Data Sources Currently Used in Life-Cycle Management

Source	Description
Ship's Maintenance Action Form (2-Kilo)	Describes equipment failure, completed maintenance, and requests for services and repaired equipment
6-Year Training Plans	Provide 6-year schedule of training activities to take place on ships
Availability and Decommissioning Schedules	Provide schedule of planned maintenance and ship decommissioning
Ballistic Missile Defense Readiness Assessments (BMDRA)	Report readiness status of ballistic missile defense systems
Battle Spares	Describes spare parts maintained aboard a ship
Casualty Reports (CASREPs)	Report equipment malfunctions or deficiencies that cannot be corrected within 48 hours and that prevent training or reduce a ship's ability to perform primary or secondary missions
Class CYBER Issues	Report on class-wide cyber problems
Class, In-Service Engineering Agent (ISEA), and Fleet Advisories	Send messages in the sailor-to-engineer (S2E) portal that advise technicians on class- or fleet-wide maintenance issues
Commonality Virtual Shelf	Provides a repository of standard architectures, design guidelines, specifications, and approved parts
Configuration Data Managers Database—Open Architecture (CDMD-OA)	Tracks status and maintenance actions of naval equipment and their related logistics items (drawings, manuals, etc.)
Current Readiness Cross Pillar Team (CRCPT) Fleet Feedback	Identifies, prioritizes, analyzes, and tracks ship-class issues
Deck Self-Assessment Groom Team (D-SAGT) Reports	Provide material condition reports from on-board training for deck system operators and maintainers
Departures from Specification	Report departures from equipment engineering requirements such as type of material, dimensions, and physical arrangements when specifications cannot be met
Diminishing Manufacturing Sources and Material Shortages (DMSMS) Cases	Report on situations when maintenance work is affected by material shortages or reduction in manufacturing sources

Table 4.1—Continued

Source	Description
Engineering Operational Sequencing System (EOSS)	Provides complete set of technically correct, properly sequenced operational and casualty control procedures for each ship type and configuration
Enterprise Remote Monitoring System (eRMS)	Provides automated collection of equipment health and status data
ISEA—Regional Maintenance Center (RMC) Collaboration Tool	Provides common area for sailors and subject matter experts (SMEs) to share advice and technical information or ask questions about maintenance tasks; often results in Technical Assistance Visit Reports (TAVRs)
INSURV Annual Report	Compiles statistical information and analysis on ships' material conditions
Maintenance Figure of Merit (MFOM)	Serves as a family of computer systems that inputs maintenance data from numerous sources and uses models developed by SMEs to output predictions about fleet readiness
Mandatory Safety Ship Change Documents (SCDs)	Records on changes to ship equipment driven by safety concerns
Navy Maintenance Database (NMD)	Monitors planning and execution of ship maintenance
New Ship Construction and Turnover Feedback	Provides obsolescence and deficiency reports (trial cards) as well as design lessons learned
Planned Maintenance System (PMS) Viewer	Provides a tool for planning, scheduling, and executing maintenance
Readiness and End-to-End Common Operating Pictures (COP)	Provide one consolidated readiness picture for tactical, operational, and strategic commands
Regional Maintenance Center (RMC)	Provide industrial repair, planning and execution, technical assistance, contract administration, and engineering services for surface-ship maintenance and modernization
Safety Investigations	Report results of investigations into major mishaps, to include material, training, and operational causes and considerations
Ship Corrective Action (SCA) Report	Documents corrective maintenance actions on ships
SUPSHIP Feedback	Provides technical feedback from new ship construction
Surface Maintenance Engineering Planning Program (SURFMEPP) Class Maintenance Plan and Technical Foundation Paper (TFP)	Provides maintenance and modernization plans; develop and issue baseline availability work packages for CNO; TFPs establish notional man-day requirements
Technical Data Management Information System (TDMIS)	Tracks relationships between equipment and technical manuals throughout their life cycle
Total Ship Readiness Assessment (TSRA)	Documents ship material conditions
United States Marine Corps (USMC) Requirements, Priorities, and Recommendations	Indicate USMC requirements that must be considered when designing new ships and altering in-service ships

Table 4.1—Continued

Source	Description
WebFLIS	Provides essential information about supply items including national stock number, manufacturers, and suppliers through a web interface connected to Federal Logistics Information Services (FLIS) data
WebSked	Provides web-enabled system for scheduling and planning movement of major maritime assets

SOURCE: NAVSEA, Overview of Life Cycle Management Group Process, October 2016.
NOTE: An example of (c) Table Superscript is to the left.

More specifically, lack of a common data source has contributed to the information disconnect between construction and sustainment life-cycle phases. For example, given the present emphasis on troubled systems with high cost per part and without being made aware of data concerning individually inexpensive but collectively costly components, PEO Ships would lack a firm basis for knowing which components most deserve attention in the design and construction process. Conversely, SEA21, being unaware of decisions being made by PEO Integrated Warfare Systems (IWS) on ship combat systems modernizations, might miss opportunities to provide feedback on existing systems that been difficult to maintain or routinely upgrade.

Feedback about troubled systems and unexpected maintenance costs could help guide new ship construction decisions. The Navy has multiple data sources that could let contractors and shipbuilders know when issues that could affect next in-class ship production crop up on a current part or ship. For example, the NAVSEA "Commonality Program integrated its Virtual Shelf (VS), a repository of standard architectures, design guidelines, specifications and approved parts list, with Norfolk Naval Shipyard's (NNSY) Standard Parts Catalog (SPC)" to help engineers "trim program development time" and reduce "total ownership cost."[1] Efforts like this, which provide feedback on existing parts in the Navy supply system (whether it is through a parts catalogue or through troubled-system reports) enable CSE purchasers and life-cycle managers to have access to the same data and are the first step to better informing design and construction decisions. However, it does not appear that the Navy is taking advantage of these opportunities.

---

[1] Joseph Battista, "Integration of Virtual Shelf and Standard Parts Catalog Saves Time and Money," Story Number: NNS140903-11, Naval Surface Warfare Center, Carderock Public Affairs, September 3, 2014.

Funding and Incentive Structures

Ship construction involves the purchase of long lead-time materials and a need to initiate long-term contracts with shipbuilders. The funding authority for Ship Construction-Navy extends for seven years from the year appropriated and encourages builders to make long-term plans for constructing a ship class. In contrast, much ship sustainment is resourced from annually appropriated operations and maintenance accounts. While funding does carry over once obligated, there is no guarantee that contracts will be renewed or that funding will be made available over the long term.

This construct is paradoxical in that it creates a long-term funding structure for the organization that is most focused on the short-term requirement of delivering a ship but imposes a short-term structure on the organizations that have the ships through the longest portion of the life cycle. Given historical execution data, the overall maintenance requirements for ships in service are in fact well known, and there is little reason to believe they will vary greatly year to year. However, year-to-year funding creates uncertainty as to which availabilities might be funded and what level of support each availability might receive. The normal congressional practice of providing annual funding through a series of continuing resolutions adds additional uncertainty, making long-term annual planning effectively impossible. The Navy routinely increases its maintenance funding during the course of a year, which, in turn, affects execution of maintenance plans and in fact increases costs.[2] This largely is the result of the characteristics of the funding source for maintenance. Having to rely on annual appropriation does not promote a long-term strategy for sustainment.

Impact of the Planning, Programming, Budgeting, and Execution System

Planning, Programming, Budgeting, and Execution (PPBE) is a DoD-wide process that has been in use since the 1960s. Its limitations and the various ways Navy organizations cope with these have been documented elsewhere.[3] However, its characteristics may exacerbate some of the issues we have noted in ship life-cycle maintenance.

PPBE's processes go on simultaneously over a period of years. Figure 4.1 depicts the phases across several fiscal years.

The major outputs from this process are annual budgets and a POM that delivers a five-year plan for funding major parts of the Navy's activities and requirements. If some activity is poorly resourced in the Program Objective Memorandum (POM), it will likely result in major adjustments in budget and might even require correction during execution, as we have indicated routinely happens in Navy ship maintenance. Figure 4.2 shows that the major input of the Navy's LCM groups activity is an input

---

[2]   Martin, McMahon, Kallimani, et al., 2017.

[3]   Irv Blickstein, John Yurchak, Bradley Martin, Jerry Sollinger, and Danny Tremblay, *Navy Planning, Programming, Budgeting and Execution*, Santa Monica, Calif.: RAND Corporation, TL-224-NAVY, 2016.

Figure 4.1
Execution of Navy Planning, Programming, Budgeting, and Execution Cycle

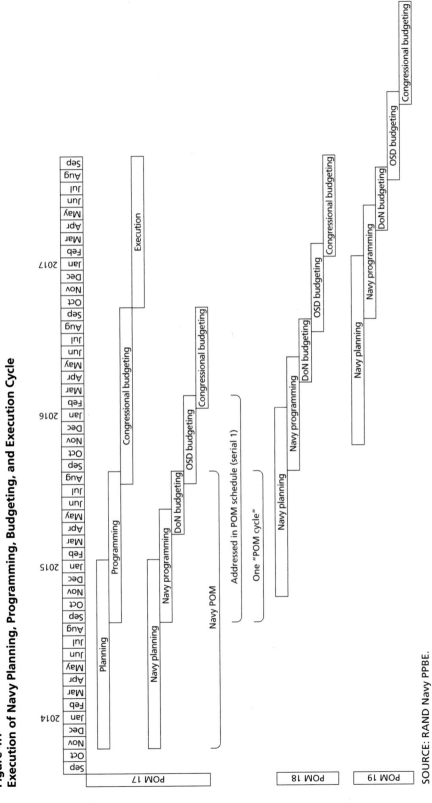

SOURCE: RAND Navy PPBE.

RAND RR2510NAVY-4.1

Figure 4.2
Life-Cycle Management Requirements Processes

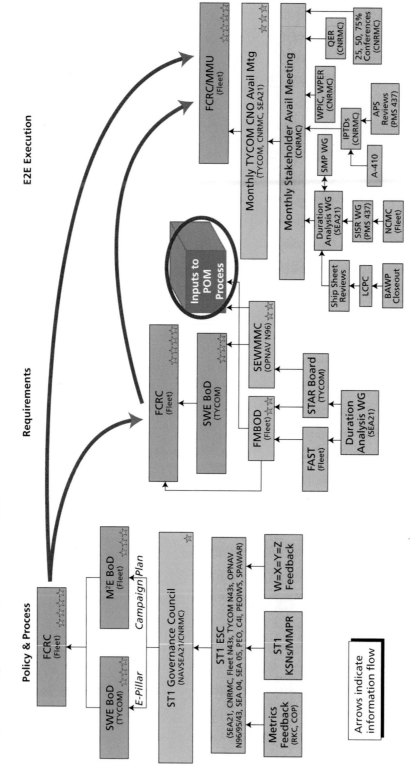

SOURCE: NAVSEA Life-Cycle Management Group (LCMG) briefing, November 2016.
NOTE: FCRC = Fleet Commander Readiness Council; SWE BoD = Surface Warfare Enterprise Board of Directors; M²E BoD = Maintenance and Modernization Enterprise Board of Directors; ST1 = Surface Team One; remainder of blocks refer to subteams and working groups for the larger parts of the organization.

RAND RR2510NAVY-4.2

into the POM. Many arrows point to the different parts of the organization and processes within them, but the one place where some request for resources comes is in the input from the requirements-generating body into the POM, which is being managed at the OPNAV level.

PPBE is a calendar-driven process, and it dominates the work of OPNAV and the organizations that provide inputs into OPNAV. Ship construction and to an even larger extent ship sustainment are on a different calendar, one driven by construction schedules, class maintenance and modernization plans, the availability of industrial facilities, operational schedules, and the unexpected events that occur in the real world. Some of the failure of issues identified in sustaining ships not being transmitted to new construction is likely due to the delay in identifying the issue as a POM input, having it placed in a sponsor's POM, and then resourced in new ship construction. At least one ship and possibly more will have been introduced into the fleet by the time the issue is actually resourced. Conversely, events that affect sustainment of ships in service happen at intervals that might not conform to a POM cycle. Delays in implementing required upgrades result because the process cannot respond any more quickly than the PPBE cycle allows.

Organizational Structure

The issues described above—lack of common data sharing and standards and disincentives generated by resource and budgeting structures—are structural in nature. The Navy could not fix the data issue on its own; it would need congressional and higher-level Executive Branch concurrence to change its method of resourcing different aspects of ship life cycles; DoD mandates the PPBE system. There may be mitigations within all of these, and we make recommendations in the following chapter.

However, broader problems exist in the way that the Navy has organized its surface-ship life-cycle enterprise. There are three separate organizations responsible for shipbuilding, maintenance, and modernization: PEO Ships, SEA21, and PEO IWS. These organizations are commanded by different flag officers, have different budget lines, and move along separate lines of effort. At the same time, under the existing enterprise construct, Commander of Naval Surface Force (CNSF) articulates the requirement while the OPNAV staff serves as the resource sponsor. Today it is not clear that the Navy's organization can meet the requirement to balance capability against cost against risk across the enterprise. In this environment, it is not clear who has the final authority. These problems diminish unity of effort and promote a stovepiped approach to life-cycle issues, a propensity that data, resource, and budget structures intensify.

Actors in the Process
Several organizations support or affect the surface LCM process. Figure 4.3 shows the current structure of NAVSEA and its affiliated PEOs.

Figure 4.3
NAVSEA and Affiliated Program Executive Office Organizational Structure

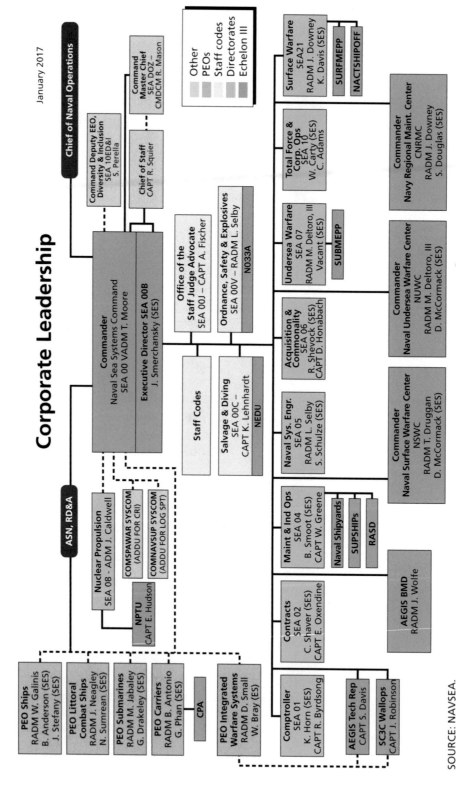

SOURCE: NAVSEA.
RAND RR2510NAVY-4.3

PEO Ships (top left side of figure) has as its stated mission the design and construction of destroyers, amphibious ships, and special mission and support ships.[4] PEO Ships reports to the ASN RD&A, although it maintains a supporting relationship with the commander of NAVSEA. The second actor is NAVSEA's Deputy Commander for Surface Warfare, SEA21 (right side of green-shaded organizations). SEA21 is the dedicated LCM organization for the Navy's in-service surface ships and is responsible for managing critical modernization, maintenance, training, and inactivation programs. According to its mission, SEA21 provides "wholeness to the Fleet by serving as the primary technical interface, ensuring surface ships are modernized with the latest technologies and remain mission relevant throughout each ship's service life."[5] SEA21 reports to the Commander of NAVSEA.

The third actor is PEO IWS (left side of diagram). Its stated mission is to "develop, deliver and sustain operationally dominant combat systems for Sailors."[6] As with PEO Ships, PEO IWS reports to ASN RD&A. PEO IWS also affects ship design decisions on the new construction side as well as modernization decisions regarding the current fleet.

The split chain of command between ASN RD&A and NAVSEA creates a budgeting stovepipe for shipbuilding/modernization, on the one hand, and fleet maintenance, on the other. Under this organizational construct, although SEA21 is responsible for managing life-cycle maintenance costs, PEO Ships and PEO IWS have locked in large portions of the life-cycle costs during the ship design phase. Additionally, throughout the life of a ship, PEO IWS–driven modernization competes with SEA21 HM&E system upgrades or new equipment installs, potentially increasing availability and reducing future maintenance costs. This is not a situation created by ill intention or lack of desire to do an effective job. However, the organization itself promotes a view that lacks cohesion.

Team Ships

In an effort to bridge the gaps between the PEO Ships and SEA21 stovepipes, Team Ships has been established, with PEO Ships and SEA21 designated as coleads. The vision for Team Ships "is to be a high performing, effective, and integrated Navy Team, providing the world's best innovation, acquisition, life-cycle-support, and disposal practices for the Navy."[7] Team Ships works to achieve this objective through multiple initiatives, working groups, and review panels (e.g., the NAVSEA Lifecycle

---

4   Multiple other stakeholders are further involved in the new-construction shipbuilding processes. These include the Supervisor of Shipbuilding Chief Engineer, the Program Manager, the Ship Design Manager, the shipbuilder, the original equipment manufacturer, the contractor, INSURV, and numerous others.

5   NAVSEA, "NAVSEA 21 Program Summary," December 2015.

6   ASECNAV, "PEO Integrated Warfare Systems," undated.

7   NAVSEA, "Team Ships," undated.

Readiness Group, the Surface Ship Revitalization Initiative Working Group, Surface-Ship Life-Cycle Management Activity, and the Fleet Review Panel).

The Team Ships concept is valid, but it does not completely overcome the natural tension that exists between the missions of the actors in the process. It is not hard to envision a case in which a design change increases cost in a current POM cycle but results in a much lower cost in the future. In the case of ship maintenance, the cost savings will likely occur outside of the Future Years Defense Plan (FYDP), and thus fall beyond a resource sponsor's maintenance planning horizon. Within the Team Ship's construct, with the two coequals reporting to two different commanders, who is responsible for conducting the life-cycle cost analysis and who makes the final call to adjudicate the issue?

Surface and Expeditionary Warfare Maintenance and Modernization Committee

In most cases, in the absence of further guidance, the resource sponsor and requirements monitor, the Directors of Surface Warfare on the CNO staff (OPNAV N96) adjudicate resource allocation during the PPBE process. N96 makes trades across the manpower, readiness, modernization, and shipbuilding accounts. In June 2016, OPNAV N9 laid out a new process to make these trades with N96 as the chair of the Surface and Expeditionary Warfare Maintenance and Modernization Committee (SEWMMC). The process used by the committee to develop the Surface and Expeditionary Warfare Maintenance and Modernization Plan (Figure 4.4) highlights the complexity and breadth of the issues that are taken into consideration. As the committee chair, N96 balances TYCOM maintenance requirements, fleet readiness requirements, and fleet maintenance priorities against OPNAV force structure requirements and capability requirements.

While this process does result in an organization taking something like a holistic view, the connection to PPBE injects rigidity and short-term focus, as we noted earlier. There is no obvious reason why OPNAV N96—generally a rear admiral—should have a better view of what is needed for long-term sustainment or near-term construction execution than SEA21 and PEO Ships, respectively. Priorities will be established; some priority decisions will be based primarily on the need to deliver a balanced POM.

Absent from the discussion in a significant way is the TYCOM and titular head of the Surface Warfare Enterprise, Commander, Naval Surface Forces (CNSF). While CNSF should have a clear view both of what's currently needed in the fleet and of what the future of the fleet should look like, in the current process CNSF's role is limited to an O-6 or equivalent Advisory Requirements Board that also reviews the input from the various action groups shown in Figure 4.2. The output of this process is the Surface and Expeditionary Warfare Maintenance & Modernization Plan that is submitted annually for the CNSF's signature. The result is a process in which the officer who certifies the plan does not own the process and whose input is limited to reviewing inputs

Figure 4.4
Surface and Expeditionary Warfare Maintenance and Modernization Plan

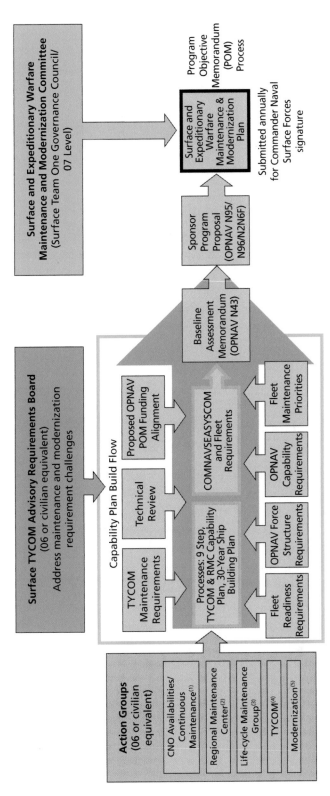

Action Group Chairs: (1) Surface Maintenance Engineering Planning Program Activity,
(2) COMNAVRMC, (3) SEA21, (4) TYCOM, (5) SEA21

SOURCE: OPNAV instruction 4700.40.

RAND RR2510NAVY-4.4

and advising the committee, which by instruction is charged to "review and incorporate surface TYCOM advisory requirement board recommendations into the surface and expeditionary warfare maintenance and modernization plan as *appropriate*." The officer who might most logically be expected to bridge the short- and long-term view may have relatively little ability to influence the process.

This structure may in fact do no harm in an era when it is widely recognized that the Navy cannot build its way to a larger fleet but in fact needs to buy down the maintenance backlog to maintain a combat-ready fleet through the end of service life. The challenge will be when the maintenance risk has been bought down and the historical trend of emphasis on the SCN accounts to provide the maximum capability at the lowest construction cost reappear.

When the historical trend reappears, the natural tendency will be to prioritize the increased capability over lowering life-cycle costs or current maintenance. This is especially true for added up-front costs that potentially save far more in the future unless there is a widely accepted business case process. In an environment where foreseeable life-cycle issues are not dealt with during new construction, the issues may not require mitigation for years after the ship enters service, which defers the problem to a different FYDP and a different account, and there is no natural advocate unless one is designated. The current structure does not deal with this issue.

Making the Case for Considering Sustainment: A Role for the Surface Maintenance Engineering Planning Program

Even if there were a common superior with the authority to order trades between construction and sustainment, making the case for a particular sustainment decision can be difficult. Some of this difficulty is due to the lack of data on long-term equipment trends. Some of it is due to the difficulty of predicting future performance to clearly demonstrate long-term benefit, especially if the performance of some new component has yet to be established in anything other than models.

However, a considerable amount of data on maintenance cost growth strongly supports avoidance of deferrals and indeed can support a case for prudent purchase of upgraded materials early in the construction process. There has been steady growth, much of it unplanned, in surface-ship availability cost, and we know with some certainty which equipment most contributes and in particular that it is most connected with corrosion.[8] The advisability of any particular investment depends on the specifics of the systems, and this study does not attempt to provide advice on specifics. However, the surface Navy does have an organization—SURFMEPP—that has existed since 2009 and has contributed to understanding patterns of failure in maintenance

---

[8] Button et al., 2015.

and integrating maintenance and modernization efforts during availabilities. It can thus provide valuable data on current ship maintenance issues, although at present, these data are used primarily to influence the processes SEA21 controls. Nonetheless, the data SURFMEPP collects and analyzes should be an integral part of decisionmaking on new ship construction and resourcing. This kind of input could contribute to production decisions on ship classes and possibly in the writing of technical specifications, and might or might not involve significant expense and additional resourcing.

CHAPTER FIVE

Conclusions and Recommendations

The actors within the surface warfare enterprise are working very hard to fix problems as they are identified. The bulk of our findings suggest that these efforts are hampered by an inability to anticipate and correct problems before they become larger. This inability is not due to a lack of effort to communicate. Rather, it is due primarily to organizational structures that impede common understanding and a coherent approach. In this chapter, we summarize our conclusions and provide recommendations that we believe will concretely address some of the identified issues.

Conclusions

These conclusions derive from the analysis we performed, our interpretation of reports generated by other researchers (in particular the GAO), and our analysis of the organizational structure as depicted and explained to us by stakeholders. There is unmistakable evidence that the process can and should be improved. The systemic problems we identified are as follows:

- **The bias of the "troubled-systems" approach toward immediate rather than long-term issues.** Troubled systems are identified after they fail in service and are heavily biased toward the individually rather than the collectively expensive item.
- **The numbers of databases and systems designed to address specific parts of the life-cycle process, coupled with the lack of coherence among them.** Even taken together, these data sources would not produce a complete and accurate picture of the state of ships in service.
- **The differing incentive structures created by differences in command focus, amplified by differences in funding mechanisms for resourcing different aspects of the ship's life cycle.**
- **The reliance on the PPBE process as a means to allocate resources.** This reliance drives a focus on influencing the POM as opposed to defining requirements and reviewing existing data. Improvement proceeds at the pace of the PPBE process and no faster.

- **The lack of a common superior empowered to adjudicate among different organizational priorities.**

All these problems together work to defeat the best efforts of the numerous organizations to anticipate rather than react to problems. Some of these issues can be dealt with directly by the Navy at the OPNAV, TYCOM, and NAVSEA levels. Others will require approval and action at higher levels but can still be mitigated by actions the Navy can take. The following recommendations are intended to promote decision coherence, common understanding of life-cycle issues, and ability to anticipate and proactively correct problems rather than react after the fact.

Recommendations

The major discrepancies in LCM are generally structural and procedural. The following recommendations are not intended to propose changes to any particular program. They are largely policy changes the evidence suggests the Navy needs to make. Where applicable, we note that some kinds of systems are available, but we do not recommend any particular system decision.

Data Sets and Systems

The first and most immediate issue, over which the Navy has more or less complete control, is generating and enforcing common data standards across the whole of the enterprise. Databases and systems were created with a good purpose, but they have proliferated and created a situation in which no clear picture can be generated of what issues most need to be corrected. Agencies see the databases they have corrected and do not always have the incentive or even the means to share information or view what information other organizations might have. To remedy this, a new data management system must include the following characteristics:

- accessibility across the enterprise
- agreement that the data derived from an agreed-upon family of systems forms the basis for decision in every aspect of ship LCM
- a common temporal horizon or at least the ability to depict the effect of decisions across multiple horizons
- ability to relate different aspects of the readiness framework in a common framework, specifically, to provide understanding of how decisions made relating to one aspect of the life cycle affects some other.

To a degree these are simply matters of policy. Navy leadership can dictate the first two conditions and direct that any system used must be available across the enterprise and that all enterprise decisions will be informed by data derived from and used by actors in the system. This recommendation does presuppose that the actors can form an idea of which data really require capture, and this understanding will require discussion on the data needs of different users rather than a simple order to implement.

Essential Common Data

The following data sets might not be comprehensive, and the intent here is not to present a final list. Rather, it is intended to suggest data about systems and processes that will span ship service life and apply to all the major elements of material readiness. This suggestion is predicated on the conclusion that the current loosely or nonfederated systems are missing important features and that this requires correction for the LCM process to be capable of proactive prevention rather than reactive correction.

- The failure rate, by component, of installed equipment across the fleet, whether this failure is corrected by repair of the component, replacement of the component, or replacement of the overall system in which it serves. This collection effort should include a crosswalk between part numbers used in the supply system and the SWLINs used in the maintenance system. The current Material Management Maintenance (3-M) System is intended to capture much of these data in the Current Ship Maintenance Project for systems known to require correction, but the quality of the inputs into this system varies widely, and it does not appear to be the case at all that 3-M data on ships in service provides any input into decisions on new ship installations. The intention is not to know where components or systems have failed but where they are planned to be installed or have been installed and not yet failed.
- The specific long-range maintenance schedules for ships in service, including the maintenance requirements identified in particular availabilities as necessary by NAVSEA Technical Warrant Holders in Technical Foundations Papers or like documents.
- The specific cost of deferred maintenance, possibly by historical comparison of jobs with like SWLINs. Some decisions on deferral would still depend on "as-found" conditions, but the normal general effect can be depicted.
- The modernization plans and schedules as generated by PEO Ships and PEO IWS, to include the maintenance availabilities in which the modifications are planned.

The Maintenance Figure of Merit System

MFOM is a family of computer systems that inputs maintenance data from numerous existing sources and uses models developed by SMEs to output predictions about fleet readiness. Variants of MFOM have been used by parts of the surface warfare enterprise since the early 2000s; the latest version was patented in 2011.[1] While it was originally

---

[1] Jarratt M. Mowery, Randy D. Bennett, Rick Leeker, and Charles W. Chesterman, Jr., "Maintenance Figure of Merit System and Method for Obtaining Material Condition of Ships," U.S. Patent Application No. 103/030,158, filed 2011.

Figure 5.1
System Inputs and Outputs

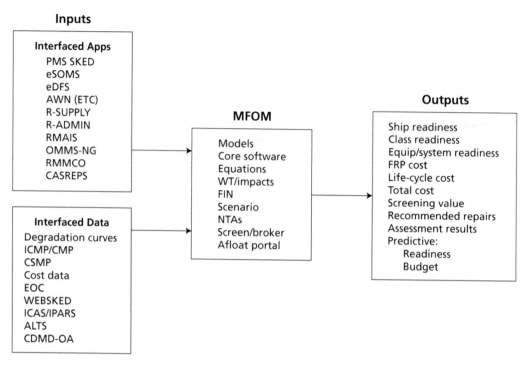

SOURCE: Mowery et al., 2011.
RAND *RR2510NAVY-5.1*

intended primarily as a means of prioritizing work between and within maintenance availabilities, it does have the potential for enabling the broader view we are recommending. Figure 5.1 shows the inputs and outputs for this system. Note that many of the inputs listed in Figure 5.1 appear in Table 4.1 as data being used by some part of the LCM enterprise. MFOM does seem to be a promising attempt at integrating pieces of the picture.

However, it is important to keep in mind the intended use for this system. Its interfaces and data are heavily oriented toward planning availabilities for ships in service. The outputs, while including life-cycle cost, primarily give recommendations about repairs to make in an availability, along with a realistic cost assessment. It takes schedule and supply data into effect, and as a result allows a more complete availability package in which requirements, schedule, and resources line up. That information is without question helpful, but the inputs might not capture systems that repeatedly create problems and thus suggest where the Navy would benefit from installing different systems earlier in the life cycle or choosing some different component. The enterprise gets a better view of what realistically correcting problems is going to cost. It might not get better insight into what could prevent the problem altogether.

The Way Ahead for Data Commonality

With an understanding of the requirements for common data sets and systems, and evidence that MFOM might help with aspects of this problem, we return to the issue of policy and organizational discipline. These data sets and systems were allowed to proliferate because the commands involved in LCM did not share a common vision and as a result focused on different things. Fixing the data problem is a necessary part of improvement, but it is not the only piece.

Incentive Structures

This is an area where the Navy cannot on its own correct the issues. Congress controls the appropriations process and has shown itself reluctant to release control over the individual accounts. There are problems resulting from an annual appropriations process for operations and maintenance, and the system would benefit from a longer-term appropriation. Navy leadership should make this case within DoD and to Congress. However, even if a change were to be implemented, it would likely be over a period of years. Hence, the Navy should probably focus its efforts on mitigating the effects of the system in the understanding that a formal charge is unlikely in the immediate future.

The major effect of existing incentive structures is that they provide little assurance that long-term planning for ship readiness will actually be rewarded with regular appropriations. Regardless of what the plan ought to be, it requires annual reappropriation, with POM and budget likely reflecting the process demand for balance rather than actual needs for sustainment. The frequent adjustment of maintenance accounts actually in execution shows that the needs—needs that might actually be reasonably well known—get underreported for the sake of process. This underreporting also contributes to a sense that the Navy is not being honest in reporting its actual needs, which has been a frequent refrain in congressional assessments of Navy readiness reporting.[2]

The Navy does place a high degree of priority on inputs into the POM process and in fact tends to measure success and failure in achieving staff objectives according to the degree to which staff actions informed and protected a POM. Removing this as a centerpiece of process and decisionmaking may be the most effective step that Navy leadership can take to counteract the incentives the system will generate if unchallenged. Clearly POM and budget submissions must occur and be balanced, but the submissions should be accompanied by a very clear statement of consequences of the trades made to achieve balance. This statement should include the cost of deferring maintenance, the specific capability upgrades foregone as a result of trades, and how near to readiness breaking points particular decisions might bring the operating force.

---

2   Rebecca Kheel, "House Panel to Hold Hearing on Navy Warship Collisions," *The Hill*, August 23, 2017.

Changing from the annual PPBE process to longer-term sustainment does not really affect organizations involved with new construction. The longer-term view and the multiyear appropriations appear to be effective. However, these need to be paired with a long-term perspective on the sustainment portion of the life cycle, even if this view has to be achieved in spite of powerful organizational predispositions.

Command and Organizational Structures

The current organizational structure for building and maintaining the surface fleet primarily divides among four different flag officers (PEO Ships, PEO IWS, SEA21, and N96) who report through three different seniors (ASN RDA, NAVSEA, and OPNAV N9). Each uses its own set of databases and funding streams (Ship Construction-Navy, Weapons Procurement-Navy, Other Procurement-Navy). Optimizing funding over requirements within their individual areas of responsibilities can and occasionally does cause an unplanned cost for one of the others. While there is no doubt that all these organizations desire cooperation to arrive at the best outcomes, the organizational structure does not naturally promote it.

Most notably, no common superior is charged with adjudicating between the programs and proposals of these organizations. Lacking such an officer, each organization will generally do what it believes is most consistent with its mission. The Navy recognizes that there is a gap between organizational priorities, and, as noted earlier, established a Surface and Expeditionary Maintenance and Modernization Committee in 2016. While this is a good start, the committee focus on in-service ships does not carry over to influence decisions being made in new construction that affect subsequent service life.

Authorities a Common Superior Would Require

Several alternatives could put a common superior in command of all these organizations, but as we consider these, we need to bear in mind that what we hope to promote is a perspective that can encompass the timelines and requirements all these organizations face. Ships must be delivered on time and within the appropriated budget; modifications that pace the threat and respond to new technological developments must be planned and executed; maintenance periods must be scheduled, and these will inevitably be affected by the needs of scheduling and operations; budget and POM submissions are not optional activities, and failure to do them properly can affect service programs.

A common superior should be able to expand the new Surface and Expeditionary Maintenance and Modernization Process to include the identification of lessons that could hold down future life-cycle costs learned from in-service ships. That officer should be able to task PEO Ships and PEO IWS to implement high payoff lessons learned into new construction ships. The Navy should structure future ship construction contracts to ensure lessons learned from in-service ships that could hold down life-cycle costs can be economically implemented into follow-on ships of the same class.

Finally, this officer should have the authority to direct budget and program priorities, to the point of being able specifically to trade off current and future readiness decisions as the situation requires. This officer and his or her organization will require an interface with the operating fleet but must also be able to represent to the CNO reasons why a short-term decision might also be the one carrying very large consequences.

Options for a Common Superior

We believe we can rule out OPNAV as being too narrowly focused on budget and POM processes. The three-star directing N9 has multiple warfare portfolios to assess within a broad fleet architecture, and considers the broad range of platforms and systems, not just surface ships. This is a needed perspective, but it is most effective in generating capability and force structure requirements. It might not be effective in trading off the details of platform choices necessary to really balance ship readiness.

We would also rule out any acquisition focused organization, such as the Deputy Assistant Secretary of the Navy for Ships (DASN-Ships). Acquisition is a complicated process with significant challenges that need to be addressed before even beginning to consider life-cycle costs. Even with direction to consider these, a shift toward considering sustainment would be major, and it is probably asking too much to expect that such an organization could easily do so.

The commander of NAVSEA would be another option. If chosen, PEOs would have to report to NAVSEA, which has never foregone responsibility for the life-cycle wholeness of ships in service. There is really little downside to this arrangement other than that NAVSEA has little direct connection with fleet requirements.

A fourth option is for the CNFS, as TYCOM, to exert greater influence in all aspects of ship design and sustainment. This would actually align with the current practices of the Naval Aviation Enterprise, and CNSF has equities in, and indeed visibilities into, the whole of the surface warfare enterprise. There is much to commend this approach, but there is one major drawback: CNSF's reporting is currently to the fleet commander, and fleet commands will generally have a bias toward short-term availability. Unless there is an organizational reason for CNSF to value longer-term sustainment, the desire to look forward may be overwhelmed by near-term demands.

We do not specifically endorse any particular common superior designation. We do strongly recommend that one be designated, with reporting lines that make clear the need to consider the short and long term.

Next Steps

These recommendations are intended to give the Navy specific choices, but with an emphasis on meeting a set of requirements rather than addressing one particular problem. The Navy will choose its own courses, but our research strongly indicates a need to select a new one. A significant proportion of the issues identified are ones that the Navy can directly control or influence, and another portion, while beyond the ability of the Navy to change on its own, are still susceptible to the Navy's influence to change its perspectives. From our vantage point, this change in perspective may be the most important next action.

Evaluation of Strategic and Comprehensive Review of Recent Surface-Force Incidents and Implications for Surface-Ship Life-Cycle Management

The U.S. Navy recently has had a number of surface accidents/incidents, some with a loss of life. In response, the Navy performed a *Comprehensive Review of Recent Surface Force Incidents*[1] and a *Strategic Readiness Review*[2] to examine the underlying causes of the incidents and to provide recommendations and actions required to address the findings.

Our sponsor asked us to review both reports to identify areas that may impact LCM of surface combatants. We first present our assessment of the *Comprehensive Review* and then the *Strategic Readiness Review*.

The LCM of surface ships is a complex endeavor with many facets. The building blocks of LCM include evaluating system/equipment metrics, performing reliability, maintainability, and maintenance availability analysis, identifying troubled systems, performing of system health assessments, conducting stakeholder reviews, obtaining fleet technical support feedback, and supporting the prioritization of fiscal investments. LCM is also responsible for integration and platform wholeness, which includes new ship construction and providing shipbuilding inputs, obsolescence management of ships' systems, stakeholder communications, management of RMCs, and technical support, as well as other responsibilities. Finally, responsibilities include funding future investments wisely.

Comprehensive Review of Recent Surface Force Incidents

Section 1 of the report, the Executive Summary, provides a summary of findings and actions as they relate to seamanship and navigational practices. We have reproduced

---

[1] Commander, Fleet Forces Command, *Comprehensive Review of Recent Surface Force Incidents*, October 26, 2017.

[2] Michael Bayer and Gary Roughead, *Strategic Readiness Review*, December 3, 2017.

the appropriate findings and actions considered to be related to LCM and posit LCM actions/impact in italicized text below.

1.3.1 Poor seamanship and failure to follow safe navigational practices

Action: Numbered Fleet Commanders establish appropriate policies for surface ships to actively transmit and use Automatic Identification System (AIS) when transiting high traffic areas.

LCM Action/Impact: Imperative that the reliability and sustainment of AIS equipment is maintained.

1.3.3 Erosion of crew readiness, planning and safety practices

Action: Conduct comprehensive Ready for Sea Assessments to determine the material and operating readiness for all Japan-based ships.

LCM Action/Impact: Support ship's hull-by-hull material condition and readiness for assessments.

Action: Develop a force generation model for ships based in Japan that addresses the increasing operational requirements, preserves sufficient maintenance and training time, and improves certification accomplishment.

LCM Action/Impact: While this action is Japan-specific, the general point is that the organizations concerned with sustainment must make a strong case for the need to conduct maintenance and modernizations on time.

1.3.5 Assessments do not reinforce effective learning

Action: Perform a baseline review of all inspection, certification, assessment, and assist visit requirements to ensure and reinforce unit readiness, unit self-sufficiency, and a culture of improvement.

LCM Action/Impact: Assess familiarity and training of the crew in the conduct of inspections and certifications, when appropriate.

1.3.7 Surface ship Bridges not modernized as an integrated control room

Action: Consolidate responsibility and authority for Bridge system modernization and improve methods for human systems integration.

Action: Establish formal policy for requalification requirements for personnel temporarily assigned to ships and when changes in equipment configuration occur.

Action: Accelerate plans to replace aging military surface search radars and electronic navigation systems.

Action: Improve stand-alone commercial radar and situational awareness piloting equipment through rapid fleet acquisition for safe navigation.

LCM Action/Impact: Wide variances in configurations exist from ship-to-ship, even within the same ship class. Gaps between the foundational training provided to enlisted crewmembers and the complexity of the technology used in modern ship control consoles make it difficult for ships' personnel to retrain and requalify for operating and technical proficiency.

Identify differences between ships, reduce variances, and/or identify and conduct training on ship-specific installations.

Section 2 of the report provides the methodology for the review, and Section 3 provides a summary and analysis of mishaps. Section 4 discusses individual training, and Section 5 addresses unit training. Section 6 provides details of Generation and Employment of Operational Forces, and Section 7 provides applicable recommendations. Section 8 addresses systemic problems, and Section 9 is an appendix summarizing recommendations. We extracted the applicable recommendations for Sections 6 and 7 and detail our assessment of the LCM actions and impact below.

6.3.1 Force Generation

6. Evaluate and recommend a maintenance and modernization scheme for all Yokosuka-based ships that takes into account the operational requirements, the training, Ship Repair Facility and industrial base capacity and make recommendations for improvement. [OPNAV/USFF/CPF/NAVSEA, 30Jun2018]

   LCM Action/Impact: Support efforts in developing a maintenance and modernization scheme for Yokosuka-based FDNF ships according to their unique operational employment.

7.3 Seamanship and Navigation Equipment Readiness and Utility

1. Consolidate responsibility and authority for Bridge system modernization and improve methods for human systems integration. Establish a single authority responsible for all Bridge system operational requirements aligned with a single engineering authority responsible to the Navy for management of the Bridge and Combat Information Center systems as an integrated control room over the life of each ship class. [OPNAV/NAVSEA/PEO IWS, 31Mar2018]

   LCM Action/Impact: Support single authority in its responsibility for management of Bridge and Combat Information Center systems as an integrated control room over the life of each ship class.

2. Accelerate plans to replace aging military surface search RADARs and electronic navigation systems. Fully fund development and implementation of Next Generation Surface Search RADAR. [NAVSEA, 31Mar2018]

    LCM Action/Impact: Replace aging surface search radars; develop next generation radars.

3. Improve stand-alone commercial RADAR and situational awareness piloting equipment through rapid fleet acquisition for safe navigation. Identify, acquire, install, and provide the maintenance and training support for an appropriately positioned common commercial RADAR system on all surface ships. Remove existing nonsupported commercial RADARs and establish policy preventing installation of commercial sensors without authorization. [NAVSEA/CNSF, 31Mar2018]

    LCM Action/Impact: Take action as indicated in support of commercial radar installations.

4. Perform a baseline review of all inspection, certification, assessment, and assist visit requirements to ensure and reinforce unit readiness, unit self-sufficiency, and a culture of improvement. The goal of this review should be to reduce the overall burden on ships by eliminating low value engagements and refocus remaining actions on validating unit readiness, unit self-sufficiency, and improvement. [CNSF/NAVSEA, 31Mar2018]

    LCM Action/Impact: Coordinate, integrate and support inspections, certifications, and assessments, and assist visits to reduce burdens on ship and to support unit readiness, self-sufficiency, and improvement.

5. Numbered Fleet Commanders establish appropriate policies for surface ships to actively transmit and use Automatic Identification Systems (AIS) when transiting high traffic areas. Consider if similar guidance is needed for submarines. [C3F/C4F/C5F/C6F/C7F/CT80, immediate]

    LCM Action/Impact: Support maintenance and operation of ships' AIS.

7. Immediately develop and release a fleet advisory on safe and effective operation for all variants of ship's steering systems in all modes of operation. [NAVSEA, immediate]

    LCM Action/Impact: Develop advisory for current ship's steering system and on a continuing basis for future modifications.

9. Accelerate plans to transition to Electronic Chart Display and Information System-Navy (ECDIS-N) versions 9.4 and greater on all ships with hardware that secures its connection and allows full AIS integration. In the interim,

NAVSEA and Type Commanders should jointly develop ship specific guidance on how to configure and operate ECDIS-N, RADARs, and AIS to maximize reliable situational awareness, reduce cyber vulnerability, and best meet the intent of International Maritime Organization standards. [NAVSEA/PEO IWS, 30Nov2017]

LCM Action/Impact: Direct action required. Take action as indicated.

10. Survey ships with integrated Bridge systems for feedback and lessons learned. [NAVSEA, 31Mar2018]

    LCM Action/Impact: Evaluate feedback and lessons learned to improve systems and communicate to fleet.

11. Conduct design and procedural reviews to reassess all variations of Bridge control systems in the fleet with focus on the complexity, suitability of human machine interfaces, and reliability of the underlying safety-critical control systems for thrust and steering. [NAVSEA, 31Mar2018]

    LCM Action/Impact: Evaluate current Bridge control system designs, modify as needed, communicate results to the fleet, and incorporate changes in new construction.

12. Assess the alignment between foundational training for enlisted operators and technicians and the technology used in integrated Bridge system consoles to ensure operators can take appropriate actions in response to equipment casualties. [NAVSEA, 31Dec2017]

    LCM Action/Impact: Support NAVSEA implementation of foundational training; identify configuration differences in integrated Bridge systems.

13. Develop standards for including human performance factors in reliability predictions for equipment modernization that increases automation. [NAVSEA, 31Mar2018]

    LCM Action/Impact: Support NAVSEA implementation.

Strategic Readiness Review

The *Strategic Readiness Review* examines the strategic factors that led to challenges in the surface force. The report starts with an Executive Summary, which includes a longitudinal discussion of degraded readiness. The chapters address operations, command and control, manning and training, fiscal disconnects, governance, and industry best practices and learning cultures, and each chapter provides recommendations. The last chapter relates to righting the ship, and we reproduce it here as overarching guidance to LCM:

> Righting the ship, to ensure the larger Navy deserves the trust of the officers and enlisted members who have put their trust in it, requires immediate, public, and sustained and committed leadership of the Secretary of the Navy and Chief of Naval Operations. Accountability must always fall primarily on commanders, but accountability must also be sought and assessed in a systemic way, at institutional levels, in the policy decisions and processes that can set the conditions for aberrant behavior and negative outcomes. Institutional accountability to reverse the multi-decade encroachment of the "normalization-of-deviation" is where the Secretary of the Navy, Chief of Naval Operations and Master Chief Petty Officer of the Navy must start.[3]

The chapter goes on to address key tenets as the Navy moves forward to reestablish readiness as a priority (including performing equipment maintenance), match supply and demand (balancing of operations and readiness), establish clear command and control relationships (reduce ambiguities), and become a true learning organization (anticipate and mitigate risks).

The report addresses broad issues related to LCM. We cull the specific recommendations that we felt most addresses LCM action required and impact.

3. Manning and Training recommendations

2. Establish a process to measure the true workload of ships' crews, both periodically and after upgrades and modernizations, to determine if manpower models adequately predict personnel requirements at sea and in port. This should include identification and quantification of added demands and additional work that affect readiness and technical qualifications.

    LCM Action/Impact: Support analysis of upgrades and modernizations on ship's crews.

4.4 Fiscal disconnect recommendations

1. Establish a better fiscal balance among the requirements for the operating tempo of the existing fleet, maintenance and material reset, required training and manning, and the resources necessary to accomplish these functions.

    LCM Action/Impact: Assert the need for maintenance resources to support the readiness of the fleet.

---

[3] Department of the Navy, Strategic Readiness Review 2017, p. 78.

4. Implement a maintenance standard that returns to longer depot maintenance periods rather than the current continuous maintenance philosophy to deal more efficiently with the impacts of emergent work and work delays.

   a. Create a means to articulate more comprehensive work packages.

   b. Reinstitute a ship-check validation process.

   LCM Action/Impact: Support NAVSEA implementation of longer depot periods and work package/ship-check validations.

Every facet of ship depot maintenance falls under the auspices of LCM and is a complex task. The report discusses ship depot maintenance, the accelerated consumption of service life of ships due to high operational demands, redundancies built into ships, standards of minimum operational equipment to support operations, and the challenges of planning and conducting maintenance availabilities. Continued NAVSEA LCM leadership is needed to support increasing demands on surface ships.

Expansion to a 355-Ship Navy and Implications for Life-Cycle Sustainment

As we were completing our study, the NAVSEA sponsor requested that we examine the implications of a buildup of the size of the Navy ship inventory from the current size of 285 to a 355-ship Navy. We first focus our efforts to examine the impact on life-cycle sustainment for large surface combatants (LSC), including the DDG-51 and CGs.

To achieve a larger fleet, some ships may need to have their service lives extended, as the shipbuilding capacity (and attrition of ships) may not be supportive of attaining the larger fleet size. In the following, we will discuss the technical maintenance requirements for the ESL of these ships, ESL of in-service LSCs today, an example approach taken by the U.K. Navy, potential impact on sustainment of a larger fleet, and considerations to support potential options to mitigate future challenges.

The plan for sustainment of ships beyond their ESL should be linked to the plan for their utilization. How does the Navy intend to use LSCs that have gone past their ESL? Some options are to continue to operate them as part of a carrier strike group (CSG), as independent deployers, or as performers of lower priority missions that free newer platforms to perform higher-end missions. The strategy and plan for their use will drive resourcing decisions for ship currency and upgrades to meet missions and threats.

Maintenance Requirements to Achieve Expected Ship Life

When ships are put in service, an ESL is established. NAVSEA examines the outfitting of the ships systems and equipment and establishes the maintenance requirements necessary to sustain the ship through its ESL. In particular, by identifying the maintenance required to meet the ESL, the TFP is the notional maintenance plan for the ship from commissioning to decommissioning. The notional long-range maintenance schedule it provides includes the man-days, availability periodicity, and availability duration for depot-level maintenance.[1]

---

[1] Naval Sea Systems Command, Surface Maintenance Engineering Planning Program (SURFMEPP), Core Products, January 2017.

With the maintenance requirements established to achieve ESL, the challenge to address is, how will an ESL extension impact the maintenance demands to sustain them beyond the established ESL for LSCs? To better explicate the expected service lives of LSCs, the paragraphs that follow will illustrate the DDG-51 and CG force structure today, duration the ships will be in service, and potential for impact of ESL extensions.

Expected Ship Life of In-Service Ships

As of this writing, there are 65 DDG-51 active ships in service[2] and 22 CGs.[3] An option to increase the size of the fleet is to keep the in-service ships in service longer. The *Annual Long-Range Plan for Construction of Naval Vessels* addresses the service life extensions (SLE) of ships. The report notes that SLEs provide near-term opportunities to sustain inventory to more rapidly achieve the Navy the Nation Needs (NNN) requirements. Because SLEs are relatively short-term extensions, they are carefully balanced with the steady long-term growth profiles to ensure overall higher (ship) numbers when SLEs expire. Candidate ships are evaluated for restoration, their ability to be upgraded with current systems, anticipated additional life, and cost vs. replacement (or other higher priority investments).[4] The need to perform required maintenance, and restore fleet ships, is imperative if the Navy is to consider SLE.

The criteria used to evaluate a candidate ship's capability for SLE, described above, includes the ability to upgrade ships with current systems. This implies that ships of a class are not configured in the same way—and some ships cannot be upgraded to current systems without major rework and resourcing. The changing nature of technology requires rapid reconfiguration to keep up with current technological advancements (and with those of potential adversaries).

An evaluation of cost of restoration, upgrading to current systems, potential for additional life, and cost versus replacement implies a cost-benefit analysis of efficacy of building new versus reconfiguring older platforms. This analysis is a complex endeavor with many variables.

The Navy's SLE criteria imply that in order to best utilize older platforms, they must have the ability to be easily restored and upgraded, have high anticipated additional life, and provide a return on investment better than a replacement (or other higher priority investments). Going forward, if the Navy is interested in the potential

---

[2]   Active ships or service craft are those that have been formally accepted by the Navy and are either in-service or in commission.

[3]   Data derived from Naval Vessel Register, NAVSEA Support Office, undated.

[4]   Deputy Chief of Naval Operations, Report to Congress on the Annual Long-Range Plan for Construction of Naval Vessels for Fiscal Year 2019, February 2018.

for maximizing the service life of new ships, these criteria will be useful in guiding designs for new ship construction.

A key attribute of DDG-51s and CGs is that they are outfitted with the Aegis Weapon System (AWS) and the Mk 41 Vertical Launching System (VLS). AWS is an advanced combat systems suite used to detect, track, and engage adversary targets. The VLS provides a unique capability to launch a variety of weapons against air, surface, and subsurface threats. The capacity of weapon load-outs and the flexibility of their employment against numerous threats make ships with VLS capability an enduring value to the fleet.

DDG-51 Class Ships

The DDG-51 class early hulls, designated Block I DDGs (DDG-51 through DDG-78) have an ESL of 35 years, while DDG-79 through the latest commissioned DDGs (Block II/IIA) have an ESL of 40 years. In addition to the active DDGs, there are nine under construction. Senior Navy leaders posit that an ESL increase of five years for DDG-51s (and CGs) is a possibility.[5] If that were to occur, an ESL extension of the DDG-51 class would increase the Block I DDGs from 35 years to 40 years, and Block II/IIA from 40 years to 45 years. Figure B.1 illustrates the number of current in-service DDG-51s in the fleet over time, as well as the how the force profile would change if ESL for each ship was pursued.

Figure B.1
Size of Current DDG-51 Class Ships Over Time, Without an Extended Service Life and With a Potential 5-year ESL

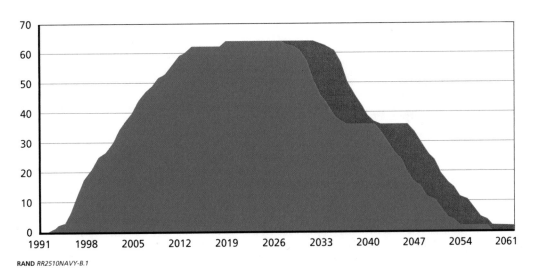

RAND RR2510NAVY-B.1

---

[5] Sydney J. Freedberg, "More Maintenance $$ Gets Navy to 355 Ships Sooner: NAVSEA," *Breaking Defense*, June 1, 2017.

An ESL extension from 35 to 40 years for Block I DDGs is a 14.3 percent life extension, and a life extension from 40 to 45 years for Block II/IIA DDGs is a 12.5 percent extension. In total, if all of the DDG-51s ESLs were extended, the result would be an additional 325 ship-years for the class. In addition, such extensions would keep the last of the (in-service) DDGs in the fleet from FY 2056 through FY 2061.

Cruisers

There are currently 22 CGs in service today, and there has been much deliberation about their future. Past debates centered on whether to put older CGs in a reduced operating status or modernize their combat systems suite. Current discussion focuses on keeping CGs in service to support a larger surface combatant fleet. For example, Commander, NAVSEA commented, "It's worth noting that Congress and the Navy have wrangled repeatedly over *retiring aging guided-missile cruisers* (CGs). Now, Moore is saying the *22 Ticonderoga-class ships* could gain at least an additional five years of life, which will please the Hill."[6]

The CG-class ship's ESL is 35 years. There are no CGs under construction, and a replacement for the CGs has yet to be determined. An ESL increase of five years for CGs would extend their service lives to 40 years. Figure B.2 illustrates the number of in-service CGs in the fleet over time, as well as the how the force profile would expand if an ESL extension for each ship was pursued.

Figure B.2
Size of Current Cruiser Fleet Over Time, Without an Extended Service Life and With a 5-Year Service Life Extension

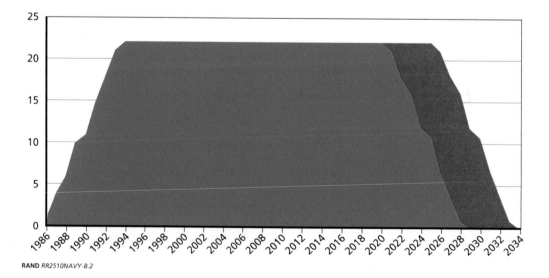

RAND RR2510NAVY-B.2

---

[6] Freedberg, 2017.

An ESL extension for CGs of 35 to 40 years is a 14.3 percent life extension. Without an ESL extension, CGs will leave the active service by 2030. With an ESL extension, the oldest CGs will remain in service through the mid-2030s. In total, if all of the CGs ESLs were extended, it would result in an additional 110 ship-years for the class. This would effectively extend the last CG in the fleet from FY 2029 through FY 2034.

Large Surface Combatant Service Life Extensions, in Aggregate

Given the discussion above, we next consider and evaluate the effect of a five-year life extension for both DDG-51s and CGs to assess the impact on LSC force structure. Figure B.3 below illustrates the current force structure and the effect of extending the service life of LSCs for five years.

The ESL extension of DDG-51 class ships and CGs would prevent a precipitous decline in LSCs and keep these capital assets in service for a combined total of 435 ship-years (335 years for DDG-51s and 110 years for CGs).

Another Navy Approach to Service Life Extensions: U.K. Navy

The United Kingdom has had experience with extending the life of its Type 23 Frigates, with ESL extensions to 32 and 35 years (from an initial anticipated service life of 18 years).

Figure B.3
Large Surface Combatant Force Structure Over Time, Without an Extended Service Life and With a 5-Year Service Life Extension

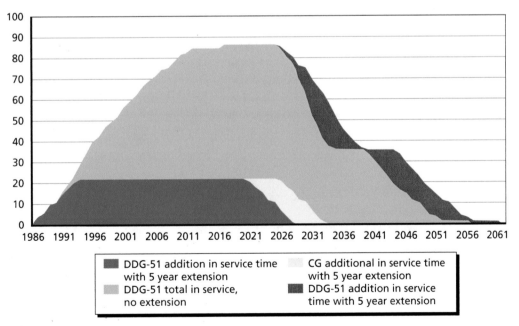

In a reply to *Jane's* regarding the announced service life extensions, the Ministry of Defense stated:

> The out-of-service dates [OSDs] for the Type 22 and Type 23 frigates have been revised a number of times in recent planning rounds. In some cases these revisions have shortened the assumed service lives of these ships, but in most cases they have extended them; extending the service of any existing ship to meet emerging requirements or to maintain overall Fleet capability until the full introduction of newer vessels, is nothing new." Indeed, maintaining a capable ship beyond its original OSD may well represent the best value option available. Such decisions are not taken lightly and the costs, benefits and risks of any extension are carefully considered in every case. In the long term, the Royal Navy is committed to an equipment programme that seeks to ensure that the ships that it operates remain among the most advanced and effective in the world and fully capable of doing all that is expected of them. . . . A consequence of the extended out-of-service dates announced this month is that the ships will now require further upgrades to prevent obsolescence in the late 2020s and early 2030s.[7]

Given this example and review of other sources, other navies extend ships' lives as a matter of course. And, indeed, the U.S. Navy has transferred (or sold) several decommissioned LSCs (including guided missile frigates [FFGs]) to foreign allied navies, some ahead of their ESL. These ships continue to be operated today even though they were deemed technologically obsolete by the U.S. Navy. This implies that these ships have useful service life remaining and can perform some defense missions, even with reduced capability.

Impact of Longer Life Cycle on Sustainment

A longer life cycle for LSCs will likely produce greater configuration challenges. While new ships in the production line are benefiting from the inclusion of the latest technology, older ships and those with SLEs may not be likely candidates for backfitting, especially if doing so is expensive. A ship approaching its ESL may or may not be a candidate for investment in its modernization. The Navy has traditionally bypassed major overhauls of soon-to-be-decommissioned ships.

There is great interest in extending ships lives, and the ability to do so exists. VADM Moore, Commander NAVSEA, noted that "from the technical side of the house, NAVSEA doesn't see anything technically that would prohibit us from extending the service life of the ships." Moreover, he added that

---

7 "UK MoD Stretches Service Lives of RN Surface Ships," *Jane's Defense Weekly*, November 21, 2008.

there's great opportunity to make the investments, a relatively small investment, to keep the ships around longer than we have today. . . . And I will point out all the time that we routinely take aircraft carriers to 50 years. And the reason we do that is because we consistently do all the maintenance you have to do on an aircraft carrier to get it to 50 years.[8]

To achieve ESL extensions, Moore notes that the primary cost would be diligently funding maintenance requirements throughout the ship's life cycle and keeping up to date with the newest combat system baselines and computer upgrades to keep the ships combat-relevant. However, in previous chapters, we note that the Navy has not funded ship maintenance accounts over time, maintenance has been deferred on LSCs, and the Navy is playing "catch-up" to address deferred maintenance backlog. In-depth reviews of individual ship's material condition and configuration need to be done to evaluate candidacy for ESL extension.

There is a point where the return on investment for ESL extensions goes south. To effectively achieve ESL extensions, ships could stay in service until the need for a dry-docking availability is presented. Dry-docking availabilities are expensive and time consuming, and the efficacy of performing this deep maintenance on older ships as compared to their utilization and capability after a docking availability may not achieve the desired results (and ties up docks that are in short supply). Therefore, ships that have been maintained, that can be cost-effectively modified to receive technologically advanced combat systems/equipment and weapons, and that do not need an extensive docking availability, would be prime candidates for ESL extension.

There are several impacts on life-cycle sustainment as the Navy grows to 355 ships. First, the maintenance industrial base must have the capacity and capability to maintain a larger fleet. The Navy must maintain not only the older and larger number of ships in service, but also newly commissioned ships in which much learning may be needed to effectively maintain and sustain them. The artisan (manpower) skills, equipment, capability, and capacity to address a new mix of ships and more of them are significant factors that will impact the sustainment of these assets. Such a major surface-ship growth enterprise needs to be assessed, planned for, and resourced for accordingly. Maintenance accounts must grow proportionately to meet this increased demand. The number of personnel who would manage a larger fleet at NAVSEA and waterfront RMCs would need to be sized appropriately with experienced staff to meet the sustainment and maintenance demands of a larger fleet. The mix of the types of ships in service will greatly affect sustainment and competition for resources, both in maintenance demands and in operations and sustainment funding. Finally, Navy docks are in short supply today, and a fleet-wide resolution of this issue must be addressed as the fleet expands, with a corresponding increase in demand for dock space.

---

[8] Megan Eckstein, "NAVSEA: Extending Surface Ship Service Lives Could Speed Up 355-Ship Buildup by 10–15 Years," U.S. Naval Institute, June 1, 2017.

There are benefits and drawbacks in keeping ships past their ESL. One benefit is that doing so eases the burden of building new ships at a high pace to meet force structure plans. Keeping ships in service longer supports a larger fleet, through reduced attrition, and a more measured pace in new ship construction. In addition, the sailors who have manned the ships and the industrial base that sustained them are current in their skills, and provide an experienced manpower pool to operate and maintain them. An increased force structure improves the Navy's ability to respond to force requirements. In addition, keeping ships in service longer maximizes use of a capital asset that is not easily or cheaply replaced. On the downside, a larger (and older) fleet will require more maintenance and sustainment resourcing. The fleet will likely have greater configuration challenges due to the length of time ships remain in service and the new technologies being built into ships in the production line. Finally, if not upgraded, they may potentially possess a reduced capability to meet new, emerging threats.

Concluding Observations

High-end LSC must be agile to meet emerging threats. Threats of today and the rapid rate of technological advancements point to the fact that future threats may be much different, and ship capabilities must be agile to meet these threats.

Based on our observations and experience, we posit that the Navy should:

Link plans for ship utilization to modernization/upgrade plans for in-service ships.

Maximize use of ships with VLSs, which provide versatility against a variety of threats. A VLS provides flexibility to load various weaponry to launch against current and potentially future threats, and offers an opportunity for reconfiguring of weapons loadout to meet perceived threats when deployed.

Resource and execute maintenance requirements throughout a ship's life cycle.

Develop designs for future ships that support modifications of new and improved equipment and/or systems and develop built in margins of space and capacity that can support configuration changes and additional/reserve power needs. In the future, it is likely that systems will demand high power beyond that generated by ship systems today.

When examining options to sustain longer ESL for new ships, identify sustainment challenges of in-service ships and address options to mitigate them in the new ship design stage.

References

"Amphibious Transport Dock—LPD: The Most Powerful Amphibious Force of All Time," *United States Navy Fact File*, May 2, 2018. As of July 13, 2018:
http://www.navy.mil/navydata/fact_display.asp?cid=4200&tid=600&ct=4

Arena, Mark, Irv Blickstein, Obaid Younossi, and Clifford A. Grammich, *Why Has the Cost of Navy Ships Risen? A Macroscopic Examination of the Trends in U.S. Naval Ship Costs over the Past Several Decades*, Santa Monica, Calif.: RAND Corporation, MG-484-NAVY, 2006. As of July 24, 2018:
https://www.rand.org/pubs/monographs/MG484.html

ASECNAV, "PEO Integrated Warfare Systems," undated. As of August 9, 2018:
http://www.secnav.navy.mil/rda/Pages/PEO_IWS.aspx

Battista, Joseph, "Integration of Virtual Shelf and Standard Parts Catalog Saves Time and Money," Story Number: NNS140903-11, Naval Surface Warfare Center, Carderock Public Affairs, September 3, 2014. As of July 13, 2018:
http://www.navy.mil/submit/display.asp?story_id=83068

Bayer, Michael, and Gary Roughead, *Strategic Readiness Review*, December 3, 2017.

Blickstein, Irv, John Yurchak, Bradley Martin, Jerry Sollinger, and Danny Tremblay, *Navy Planning, Programming, Budgeting and Execution*, Santa Monica, Calif.: RAND Corporation, TL-224-NAVY, 2016. As of July 24, 2018:
https://www.rand.org/pubs/tools/TL224.html

Button, Robert, Bradley Martin, Jerry Sollinger, and Abraham Tidwell, *Assessment of Surface Ship Maintenance Requirements*, Santa Monica, Calif.: RAND Corporation, RR-1155-NAVY, 2015. As of July 24, 2018:
https://www.rand.org/pubs/research_reports/RR1155.html

Commander, Fleet Forces Command, *Comprehensive Review of Recent Surface Force Incidents*, October 26, 2017.

Defense Acquisition University, "Defense Acquisition Life Cycle Wall Chart," February 14, 2018. As of July 13, 2018:
https://www.dau.mil/tools/t/Department-of-Defense-Acquisition-Life-Cycle-Chart

Department of the Navy, Strategic Readiness Review 2017, p. 78.

Deputy Chief of Naval Operations, Report to Congress on the Annual Long-Range Plan for Construction of Naval Vessels for Fiscal Year 2019, February 2018.

Eckstein, Megan, "NAVSEA: Extending Surface Ship Service Lives Could Speed Up 355-Ship Buildup by 10–15 Years," U.S. Naval Institute, June 1, 2017.

Freedberg, Sydney J., "LCS Troubles May Stem from Double Engine," *Breaking Defense*, September 7, 2016. As of July 12, 2018:
https://breakingdefense.com/2016/09/lcs-troubles-may-stem-from-double-engine/

———, "More Maintenance $$ Gets Navy to 355 Ships Sooner: NAVSEA," *Breaking Defense*, June 1, 2017. As of July 13, 2018:
https://breakingdefense.com/2017/06/more-maintenance-gets-navy-to-355-ships-sooner-navsea/

FY 15 Navy Programs, "Littoral Combat Ship (LCS) and Associated Mission Modules," undated. As of July 13, 2018:
http://www.dote.osd.mil/pub/reports/FY2015/pdf/navy/2015lcs.pdf

Gady, Franz-Stefan, "Dropping Like Flies: Third US Navy Littoral Combat Ship Out of Action," *The Diplomat*, August 13, 2016. As of July 13, 2018:
https://thediplomat.com/2016/08/dropping-like-flies-third-us-navy-littoral-combat-ship-out-of-action/

GAO—see United States Government Accountability Office.

Howard, Michelle J., *Command Investigation of Diesel Engine and Related Maintenance and Quality Assurance Issues Aboard USS* San Antonio *(LPD17)*, Norfolk, Va., Department of the Navy, January 15, 2010.

Kheel, Rebecca, "House Panel to Hold Hearing on Navy Warship Collisions," *The Hill*, August 23, 2017. As of July 13, 2018:
http://thehill.com/policy/defense/347695-house-panel-to-examine-warship-collisions

LaGrone, Sam, "Raytheon Excalibur Round Set to Replace LRLAP on Zumwalts," *USNI News*, January 1, 2017. As of July 13, 2018:
https://news.usni.org/2016/12/13/raytheon-excalibur-round-set-replace-lrlap-zumwalts

Martin, Bradley, and Michael E. McMahon, *Future Aircraft Carrier Options*, Santa Monica, Calif.: RAND Corporation, RR-2006-NAVY, 2017. As of July 26, 2018:
https://www.rand.org/pubs/research_reports/RR2006.html

Martin, Bradley, Michael E. McMahon, James G. Kallimani, and Tim Conley, *Accounting for Growth in the Ship Depot Maintenance Account*, Santa Monica, Calif.: RAND Corporation, RR-1837-NAVY, 2017. As of July 24, 2018:
https://www.rand.org/pubs/research_reports/RR155.html

Martin, Bradley, Michael E. McMahon, Jessie Riposo, James G. Kallimani, Angelena Bohman, Alyssa Ramos, and Abby Schendt, *A Strategic Assessment of the Future of U.S. Navy Ship Maintenance: Challenges and Opportunities*, Santa Monica, Calif.: RAND Corporation, RR-1951-NAVY, 2017. As of July 24, 2018:
https://www.rand.org/pubs/research_reports/RR1951.html

Mizokami, Kyle, "The USS *Zumwalt* Can't Fire Its Guns Because the Ammo Is Too Expensive," *Popular Mechanics*, November 7, 2016. As of July 13, 2018:
http://www.popularmechanics.com/military/navy-ships/a23738/uss-zumwalt-ammo-too-expensive/

Mowery, Jarratt M., Randy D. Bennett, Rick Leeker, and Charles W. Chesterman, Jr., "Maintenance Figure of Merit System and Method for Obtaining Material Condition of Ships," U.S. Patent Application No. 103/030,158, filed 2011.

Naval Sea Systems Command, *The Navy Ship Design Process*, Naval Surface Warfare Center-Carderock, Bethesda, Md., January 6, 2012.

———, Supervisor of Shipbuilding, "Engineering and Technical Oversight," *SUPSHIP Operations Manual (SOM)*, Chapter 8, S0300-B2-MAN-010, Bethesda, Md., August 28, 2015.

————, Surface Maintenance Engineering Planning Program (SURFMEPP), Core Products, January 2017. As of July 13, 2018:
http://www.navsea.navy.mil/Home/Team-Ships/NAVSEA-21/SURFMEPP/Mission-Statement/

Naval Sea Systems Command, "NAVSEA 21 Program Summary," December 2015. As of August 9, 2018:
http://www.navsea.navy.mil/Home/Team-Ships/NAVSEA-21/

————, "Team Ships," undated. As of August 9, 2018:
http://www.navsea.navy.mil/Home/Team-Ships/

Naval Vessel Register, NAVSEA Support Office, undated. As of July 13, 2018:
http://www.nvr.navy.mil/SHIPS.html

NAVSEA—see Naval Sea Systems Command

NAVSEA Life-Cycle Management Group (LCMG) briefing, November 2016.

Office of the Secretary of Defense, Cost Assessment and Program Evaluation, *O&S Cost Estimating Guide*, March 2014.

Pennington, Tim, "Navy Is Full Steam Ahead on Powder Coating," *PF: Products Finishing*, February 1, 2012. As of July 13, 2018:
https://www.pfonline.com/articles/navy-is-full-steam-ahead-on-powder-coating

"UK MoD Stretches Service Lives of RN Surface Ships," *Jane's Defense Weekly*, November 21, 2008. As of July 13, 2018:
http://janes.ihs.com/DefenceEquipment/Display/1178834

U.S. Government Accountability Office, "Navy Actions Needed to Optimize Ship Crew Size and Reduce Total Ownership Costs," Washington, D.C.: GAO-03-20, June 2003.

————, *Navy Shipbuilding: Opportunities Exist to Improve Practices Affecting Quality*, Washington, D.C.: GAO-14-122, November 19, 2013.

————, "Navy Shipbuilding: Policy Changes Needed to Improve the Post-Delivery Process and Ship Quality," Report to Congressional Committee on Armed Services, U.S. Senate, Washington, D.C.: GAO-17-418, July 2017. As of November 14, 2017:
https://www.gao.gov/assets/690/685799.pdf